Studies in Fuzziness and Soft Computing

Volume 328

Series editor

Janusz Kacprzyk, Polish Academy of Sciences, Warsaw, Poland
e-mail: kacprzyk@ibspan.waw.pl

About this Series

The series "Studies in Fuzziness and Soft Computing" contains publications on various topics in the area of soft computing, which include fuzzy sets, rough sets, neural networks, evolutionary computation, probabilistic and evidential reasoning, multi-valued logic, and related fields. The publications within "Studies in Fuzziness and Soft Computing" are primarily monographs and edited volumes. They cover significant recent developments in the field, both of a foundational and applicable character. An important feature of the series is its short publication time and world-wide distribution. This permits a rapid and broad dissemination of research results.

More information about this series at http://www.springer.com/series/2941

Deng-Feng Li

Linear Programming Models and Methods of Matrix Games with Payoffs of Triangular Fuzzy Numbers

Springer

Deng-Feng Li
School of Economics and Management
Fuzhou University
Fuzhou
China

ISSN 1434-9922 ISSN 1860-0808 (electronic)
Studies in Fuzziness and Soft Computing
ISBN 978-3-662-48474-6 ISBN 978-3-662-48476-0 (eBook)
DOI 10.1007/978-3-662-48476-0

Library of Congress Control Number: 2015951775

Springer Heidelberg New York Dordrecht London

Printed on acid-free paper

Springer-Verlag GmbH Berlin Heidelberg is part of Springer Science+Business Media
(www.springer.com)

To my wife, Wei Fei and To my son,
Wei-Long Li

Preface

Two-person zero-sum finite games, which often are called matrix games for short, are an important part of noncooperative games. Matrix games have been extensively studied and successfully applied to many fields such as management science, decision science, operational research, economics, finance, business, social science, and biology as well as engineering. However, the assumption that all payoffs are precise common knowledge to the players is not realistic in many competitive or antagonistic decision occasions. In fact, more often than not, in real competitive or antagonistic situations, the players cannot exactly estimate payoffs in the game due to lack of adequate information and/or imprecision of the available information on the environment. This lack of precision and certainty may be appropriately modeled by using the fuzzy set. Intervals and triangular fuzzy numbers, which are special and simple cases of the fuzzy sets, seem to be suitable and convenient for dealing with fuzziness or imprecision of payoffs in matrix games. On the other hand, in some real-life game problems, choice of strategies for the players is constrained due to some practical reason why this should be, i.e., not all mixed (or pure) strategies in a game are permitted for each player. As a result, there appear four important types of matrix games, which are interval-valued matrix games, matrix games with payoffs of triangular fuzzy numbers, interval-valued constrained matrix games, and constrained matrix games with payoffs of triangular fuzzy numbers. As far as I know, however, there is less investigation on them. Therefore, this book focuses on studying the concepts, properties, models, and methods of the aforementioned four types of matrix games.

This book is divided into two parts. Each part includes two chapters. Chapter 1 discusses interval-valued matrix games, mainly including interval-valued mathematical programming models of interval-valued matrix games, acceptability-degree-based linear programming models and method of interval-valued matrix games, the lexicographic method of interval-valued matrix games, and primal-dual linear programming models and method of interval-valued matrix games. Chapter 2 studies matrix games with payoffs of triangular fuzzy numbers, mainly including fuzzy multi-objective programming models and fuzzy linear programming method of matrix games with payoffs of triangular fuzzy numbers, two-level linear programming models and method of matrix games with payoffs of triangular fuzzy numbers, the lexicographic method of matrix games with payoffs of triangular fuzzy numbers, and Alfa-cut-based primal-dual linear programming models and method of matrix games with payoffs of triangular fuzzy numbers. Chapter 3 expatiates interval-valued constrained matrix games, including the concepts of solutions of interval-valued constrained matrix games and properties, and primal-dual linear programming models and method of interval-valued constrained matrix games. Chapter 4 expounds constrained matrix games with payoffs of triangular fuzzy numbers, mainly including fuzzy multi-objective programming models and method of constrained matrix games with payoffs of triangular fuzzy numbers, and Alfa-cut-based primal-dual linear programming models and method of constrained matrix games with payoffs of triangular fuzzy numbers. The aim of this book was to develop and establish simple, efficient, and effective linear programming models and methods for solving interval-valued matrix games, interval-valued constrained matrix games, matrix games with payoffs of triangular fuzzy numbers, and constrained matrix games with payoffs of triangular fuzzy numbers. I tried my best to ensure that the models and methods developed in this book are of practicability, maneuverability, and universality.

This book is addressed to people in theoretical researches and practical applications from different fields and disciplines such as decision science, game theory, management science, fuzzy sets or fuzzy mathematics, applied mathematics, optimizing design of engineering and industrial system, expert system, and social economy as well as artificial intelligence. Moreover, it is also addressed to teachers, postgraduates, and doctors in colleges and universities in different disciplines or majors: decision analysis, management, operation research, fuzzy mathematics, fuzzy system analysis, systems engineering, project management, industrial engineering, hydrology, and water resources.

First of all, special thanks are due to my coauthor Chun-Tian Cheng and my doctoral graduates Jiang-Xia Nan, Fang-Xuan Hong for completing and publishing several articles. This book was supported by the Key Program of the National Natural Science Foundation of China (No. 71231003), the National Natural Science Foundation of China (Nos. 71171055, 71101033, and 71001015), the "Chang-Jiang Scholars" Program (the Ministry of Education of China), the Program for New Century Excellent Talents in University (the Ministry of Education of China, NCET-10-0020), and the Specialized Research Fund for the Doctoral Program of Higher Education of China (No. 20113514110009) as well as "Science and Technology Innovation Team Cultivation Plan of Colleges and Universities in Fujian Province." I would like to acknowledge the encouragement and support of my wife as well as the understanding of my son.

Last but not least, I would like to acknowledge the encouragement and support of all my friends and colleagues.

Ultimately, I should claim that I am fully responsible for all errors and omissions in this book.

August 2015 Deng-Feng Li

Contents

About the Author

Deng-Feng Li was born in 1965. He received the B.Sc. and M.Sc. degrees in applied mathematics from the National University of Defense Technology, Changsha, China, in 1987 and 1990, respectively, and the Ph.D. degree in system science and optimization from the Dalian University of Technology, Dalian, China, in 1995.

From 2003 to 2004, he was a Visiting Scholar with the School of Management, University of Manchester Institute of Science and Technology, Manchester, UK. He is currently a Distinguished Professor of "Chang-Jiang Scholars" Program, Ministry of Education of China and "Min-Jiang Scholarship" Distinguished Professor with the School of Economics and Management, Fuzhou University, Fuzhou, China. He has been conferred the Outstanding Contribution Experts of the National Middle-Aged and Young of China and was approved as an expert of the Enjoyment of the State Council Special Allowance of China. He has authored or coauthored more than 300 journal papers and seven monographs. He has coedited one proceeding of the international conference and two special issues of journals and won 23 academic achievements and awards such as Chinese State Natural Science Award and 2013 IEEE Computational Intelligence Society IEEE Transactions on Fuzzy Systems Outstanding paper award. His current research interests include classical and fuzzy game theory, fuzzy decision analysis, group decision making, supply chain, fuzzy sets and system analysis, fuzzy optimization, and differential game. He is the Editor-in-chief of *International Journal of Fuzzy System Applications* and Associate Editors and/or Editors of several international journals.

Abstract

This book is an academic monograph based on the papers published in international famous journals by the author. The focus of this book is on theoretical models and methods of interval-valued constrained and unconstrained matrix games, and constrained and unconstrained matrix games with payoffs of triangular fuzzy numbers. This book includes four chapters. Chapter 1 mainly discusses interval-valued mathematical programming models of interval-valued matrix games, acceptability-degree-based linear programming models and method of interval-valued matrix games, the lexicographic method of interval-valued matrix games, and primal-dual linear programming models and method of interval-valued matrix games. Chapter 2 mainly studies fuzzy multi-objective programming models and fuzzy linear programming method of matrix games with payoffs of triangular fuzzy numbers, two-level linear programming models and method of matrix games with payoffs of triangular fuzzy numbers, the lexicographic method of matrix games with payoffs of triangular fuzzy numbers, and Alfa-cut-based primal-dual linear programming models and method of matrix games with payoffs of triangular fuzzy numbers. Chapter 3 expatiates the concepts of solutions of interval-valued constrained matrix games and properties, and primal-dual linear programming models and method of interval-valued constrained matrix games. Chapter 4 mainly expounds fuzzy multi-objective programming models and method of constrained matrix games with payoffs of triangular fuzzy numbers, and Alfa-cut-based primal-dual linear programming models and method of constrained matrix games with payoffs of triangular fuzzy numbers. The aim of this book was to develop and establish simple, efficient, and effective linear programming models and methods for solving interval-valued matrix games, interval-valued constrained matrix games, matrix games with payoffs of triangular fuzzy numbers, and constrained matrix games with payoffs of triangular fuzzy numbers.

This book is addressed to people in theoretical researches and practical applications from different fields and disciplines such as decision science, game theory, management science, fuzzy sets or fuzzy mathematics, applied mathematics, optimizing design of engineering and industrial system, expert system, and social economy as well as artificial intelligence. Moreover, it is also addressed to teachers, postgraduates, and doctors in colleges and universities in different disciplines or majors: decision analysis, management, operation research, fuzzy mathematics, fuzzy system analysis, systems engineering, project management, industrial engineering, applied mathematics, hydrology, and water resources.

Part I
Models and Methods of Matrix Games with Payoffs of Triangular Fuzzy Numbers

Chapter 1
Interval-Valued Matrix Games

1.1 Introduction

Game theory is engaged in competing and strategic interaction among players in management science, operational research, economics, finance, business, social science, biology, engineering, and others. It began in the 1920s and has achieved a great success [1, 2]. The simplest game is the zero-sum game involving only two players with finite pure strategies (i.e., options), which is often called the matrix game for short. A matrix game is usually expressed by a payoff matrix $\boldsymbol{A} = (a_{ij})_{m \times n}$, where a_{ij} is the amount of reward/loss which the player I wins (and hereby the player II loses) when the players I and II choose their pure strategies $\delta_i \, (i = 1, 2, \ldots, m)$ and $\beta_j \, (j = 1, 2, \ldots, n)$, respectively. Here, m and n are two arbitrary positive integers.

Traditionally, the payoffs a_{ij} are represented by crisp values, which indicate that they are precisely known. This assumption is reasonable for clearly defined games, which have many useful applications, especially in finance, management and decision making systems [3, 4]. In the real world, however, there are some cases in which the payoffs are not fixed numbers known and have to be estimated even though two players do not change their strategies. An example is one in which different advertising strategies of two competing companies lead to different market shares and the market shares must be estimated. Hence, fuzzy games have been extensively studied. Dubois and Prade [5] gave a brief overview and discussion on the fuzzy games with crisp sets of strategies and fuzzy payoffs due to the lack of precision on the knowledge of the associated payoffs. Nishizaki and Sakawa [2] and Bector and Chandra [4] made good overviews on update research of this topic. Bector et al. [6, 7] studied the matrix games with fuzzy goals and fuzzy payoffs by using the defined fuzzy linear programming duality, respectively. Campos [8], Campos and Gonzalez [9] and Campos et al. [10] proposed ranking function based

D.-F. Li, *Linear Programming Models and Methods of Matrix Games with Payoffs of Triangular Fuzzy Numbers*, Studies in Fuzziness and Soft Computing 328 ,
DOI 10.1007/978-3-662-48476-0_1

methods for solving fuzzy matrix games. However, only crisp solutions were provided [8–10]. Maeda [11] defined the equilibrium strategy of fuzzy matrix games by using the fuzzy max order. Again, only crisp solutions were provided. Also the classical minimax theorems [12] were not utilized. One theoretically sound property of game theory is that the mathematical models of the matrix game formulated from the standpoints of the two players are a pair of linear programming models which are dual of each other. Hence, solving either of the linear programming models can obtain the strategies of the two players by applying the duality theorem of linear programming. Nishizaki and Sakawa [13, 14] proposed fuzzy linear programming models of fuzzy matrix games, which only provided crisp solutions. In most of the fuzzy matrix games, the payoffs were viewed as fuzzy numbers and assumed that their membership functions are already known a priori. These membership functions play an important role in corresponding methods [15, 16]. In reality, it is not always easy for the players to specify the membership functions in fuzzy environments [17]. In some cases, the payoffs may only vary within certain ranges for fixed strategies and may be considered as interval estimates, i.e., the matrix games with payoffs represented by intervals [12, 18, 19], which usually are called the interval-valued matrix games for short. Interval computing has been a well established field by Moore [20] and successfully applied to some areas. Also interval-valued linear programming problems have been studied in details and duality results have been obtained [21–23]. Stated as earlier, the matrix game is mathematically equivalent to a pair of primal-dual linear programming problems [1, 3]. Thus, theoretically interval-valued matrix games are solvable by using the interval-valued linear programming methodology. Recently, Collins and Hu [24] investigated crisply and fuzzily determined interval-valued matrix games by using an appropriate fuzzy interval comparison operator and theoretically proposed a pair of interval-valued linear programming models for both the players. In order to be able to transform these models into the classical standard linear programming models, Collins and Hu [24] assumed that the player I's gain-floor and the player II's loss-ceiling are trivial intervals, i.e., real numbers. As pointed out by Collins and Hu themselves [24], this assumption seems to be unrealistic and unreasonable in that the value of the interval-valued matrix game being a linear combination of the entries in the interval-valued payoff matrix should be an interval from the viewpoint of a logic. It is worthwhile pointing out that Collins and Hu [24] further proposed an important technique for solving generic interval-valued inequalities through introducing the interval comparison operator or fuzzy ranking index, which has a good potential of application to the interval-valued matrix games. Liu and Kao [18] estimated the upper and lower bounds of the value of the interval-valued matrix game through developing a pair of two-level mathematical programming models, which were transformed into a pair of ordinary one-level linear programming models by the duality theorem of linear programming and a variable substitution technique. But Liu and Kao [18] focused on how to obtain the lower and upper bounds of the value of the interval-valued matrix game and did not propose any specific method for solving corresponding optimal strategies for the players. In addition, the method [18] resulted in many additional variables and constraints in the derived auxiliary linear programming models, which require a large amount of computation. Nayak and Pal [25] constructed a pair of interval-valued linear programming models for the interval-valued matrix game. However, Nayak and Pal [25] chose only the lower bounds of the player I's gain-floor and the player II's loss-ceiling as objective functions and hereby transformed the interval-valued linear programming models into the classical linear programming models in terms of the interval inequality relations [22, 23]. The resulting inappropriate formulations and vital mistakes have been pointed out and corrected by Li [26]. The bi-objective linear programming models were derived and suggested to be solved by the lexicographic method [27]. Based on the defined interval inequality relations and the fuzzy ranking index, Li et al. [28] derived a pair of bi-objective linear programming models from the constructed auxiliary interval-valued programming models for the interval-valued matrix game. The bi-objective linear programming models were solved by using the weighted average method rather than the lexicographic method [27]. Essentially, the weighted average method [27] is a ranking one based on the acceptability index of the interval comparison operator [22, 23].

Stated as earlier, the value of the interval-valued game matrix should be an interval from the viewpoint of logic. Thus, this chapter focuses on developing some simple and effective linear programming methods for solving any interval-valued game matrix.

1.2 Matrix Games and Auxiliary Linear Programming Models

Assume that $S_1 = \{\delta_1, \delta_2, \ldots, \delta_m\}$ and $S_2 = \{\beta_1, \beta_2, \ldots, \beta_n\}$ are sets of pure strategies for two players I and II, respectively. A payoff matrix for the player I is expressed with

$$\boldsymbol{A} = (a_{ij})_{m\times n} = \begin{array}{c} \\ \delta_1 \\ \delta_2 \\ \vdots \\ \delta_m \end{array} \begin{array}{c} \begin{array}{cccc} \beta_1 & \beta_2 & \cdots & \beta_n \end{array} \\ \begin{pmatrix} a_{11} & a_{12} & \cdots & a_{1n} \\ a_{21} & a_{22} & \cdots & a_{2n} \\ \vdots & \vdots & \cdots & \vdots \\ a_{m1} & a_{m2} & \cdots & a_{mn} \end{pmatrix} \end{array}$$

and hereby the payoff matrix for the player II is equal to

$$-\boldsymbol{A} = (-a_{ij})_{m\times n} = \begin{matrix} \\ \delta_1 \\ \delta_2 \\ \vdots \\ \delta_m \end{matrix} \begin{matrix} \beta_1 & \beta_2 & \cdots & \beta_n \\ \left(\begin{matrix} -a_{11} \\ -a_{21} \\ \vdots \\ -a_{m1} \end{matrix} \right. & \begin{matrix} -a_{12} \\ -a_{22} \\ \vdots \\ -a_{m2} \end{matrix} & \begin{matrix} \cdots \\ \cdots \\ \cdots \\ \cdots \end{matrix} & \left. \begin{matrix} -a_{1n} \\ -a_{2n} \\ \vdots \\ -a_{mn} \end{matrix} \right) \end{matrix}.$$

It is customary to assume that the player I is a maximizing player and the player II is a minimizing player. The vectors $\boldsymbol{y} = (y_1, y_2, \ldots, y_m)^{\mathrm{T}}$ and $\boldsymbol{z} = (z_1, z_2, \ldots, z_n)^{\mathrm{T}}$ are mixed strategies for the players I and II, respectively, where $y_i\ (i = 1, 2, \ldots, m)$ and $z_j\ (j = 1, 2, \ldots, n)$ are probabilities in which the players I and II choose their pure strategies $\delta_i \in S_1\ (i = 1, 2, \ldots, m)$ and $\beta_j \in S_2\ (j = 1, 2, \ldots, n)$, respectively; the symbol "T" is the transpose of a vector/matrix. A pure strategy $\delta_i \in S_1$ (or $\beta_j \in S_2$) may be regarded as a special case of the mixed strategy $\boldsymbol{y}$ (or $\boldsymbol{z}$), i.e., $y_i = 1$ and $y_k = 0\,(k = 1, 2, \ldots, m;\ k \neq i)$ (or $z_j = 1$ and $z_l = 0\,(l = 1, 2, \ldots, n;\ l \neq j)$). Sets of mixed strategies for the players I and II are denoted by Y and Z, respectively, i.e., $Y = \{\boldsymbol{y} | \sum_{i=1}^{m} y_i = 1, y_i \geq 0\ (i = 1, 2, \ldots, m)\}$ and $Z = \{\boldsymbol{z} | \sum_{j=1}^{n} z_j = 1, z_j \geq 0\ (j = 1, 2, \ldots, n)\}$. Thus, a two-person zero-sum finite matrix game may be expressed with the triplet $G = \{Y, Z, \boldsymbol{A}\}$. In the sequent, such a two-person zero-sum finite matrix game usually is simply called as a matrix game $\boldsymbol{A}$.

Suppose that the players I and II are playing the matrix game $\boldsymbol{A}$. If the player I chooses a mixed strategy $\boldsymbol{y} \in Y$ and II chooses a mixed strategy $\boldsymbol{z} \in Z$, then the player I's expected payoff can be computed as follows:

$$\boldsymbol{y}^{\mathrm{T}}\boldsymbol{A}\boldsymbol{z} = \sum_{i=1}^{m}\sum_{j=1}^{n} y_i a_{ij} z_j. \tag{1.1}$$

Assume that the player I uses any mixed strategy $\boldsymbol{y} \in Y$. Then, the player I's expected gain-floor is obtained as follows:

$$\upsilon(\boldsymbol{y}) = \min_{\boldsymbol{z}\in Z}\{\boldsymbol{y}^{\mathrm{T}}\boldsymbol{A}\boldsymbol{z}\}. \tag{1.2}$$

Here, $\boldsymbol{y}^{\mathrm{T}}\boldsymbol{A}\boldsymbol{z}$ can be thought of as a weighted average of the expected payoffs for the player I if he/she uses $\boldsymbol{y} \in Y$ against the player II's pure strategies. Thus, the minimum is attained by some pure strategy $\beta_j \in S_2$ as follows:

$$\upsilon(\boldsymbol{y}) = \min_{1\leq j\leq n}\{\boldsymbol{y}^{\mathrm{T}}\boldsymbol{A}_{\cdot j}\} = \min_{1\leq j\leq n}\{\sum_{i=1}^{m} y_i a_{ij}\}, \tag{1.3}$$

where $\boldsymbol{A}_{\cdot j}$ is the jth column of the payoff matrix $\boldsymbol{A}$, i.e., $\boldsymbol{A}_{\cdot j} = (a_{1j}, a_{2j}, \ldots, a_{mj})^{\mathrm{T}}$. Hence, the player I should choose $\boldsymbol{y}^* \in Y$ so as to maximize $\upsilon(\boldsymbol{y})$, i.e., so as to obtain

$$v = \upsilon(\boldsymbol{y}^*) = \max_{\boldsymbol{y}\in Y}\{\upsilon(\boldsymbol{y})\} = \max_{\boldsymbol{y}\in Y}\min_{1\le j\le n}\{\boldsymbol{y}^{\mathrm{T}}\boldsymbol{A}_{\cdot j}\}. \tag{1.4}$$

Such $\boldsymbol{y}^* \in Y$ is called the player I's maximin (or optimal) strategy, and $\upsilon(\boldsymbol{y}^*)$ is called the value of the matrix game $\boldsymbol{A}$ for the player I, denoted by $v = \upsilon(\boldsymbol{y}^*)$.

It is easy to see that computing an optimal strategy $\boldsymbol{y}^* \in Y$ and the value $v = \upsilon(\boldsymbol{y}^*)$ for the player I is equivalent to solving the linear programming model as follows:

$$\max\{\upsilon\}$$
$$\text{s.t.}\begin{cases} \sum\limits_{i=1}^{m} a_{ij}y_i \ge \upsilon \quad (j = 1, 2, \ldots, n) \\ \sum\limits_{i=1}^{m} y_i = 1 \\ y_i \ge 0 \quad (i = 1, 2, \ldots, m) \\ \upsilon \text{ unrestricted in sign.} \end{cases} \tag{1.5}$$

Similarly, if the player II chooses any mixed strategy $z \in Z$, then he/she obtains the expected loss-ceiling as follows:

$$\omega(z) = \max_{1\le i\le m}\{\boldsymbol{A}_{i\cdot}z\} = \max_{1\le i\le m}\{\sum_{j=1}^{n} a_{ij}z_j\}, \tag{1.6}$$

where $\boldsymbol{A}_{i\cdot}$ is the i-th row of the payoff matrix $\boldsymbol{A}$, i.e., $\boldsymbol{A}_{i\cdot} = (a_{i1}, a_{i2}, \ldots, a_{in})$. Hence, the player II should choose $z^* \in Z$ so as to obtain

$$\mu = \omega(z^*) = \min_{z\in Z}\{\omega(z)\} = \min_{z\in Z}\max_{1\le i\le m}\{\boldsymbol{A}_{i\cdot}z\}. \tag{1.7}$$

Such $z^* \in Z$ is called the player II's minimax (or optimal) strategy, and $\omega(z^*)$ is called the value of the matrix game $\boldsymbol{A}$ for the player II, denoted by $\mu = \omega(z^*)$.

Obviously, computing an optimal strategy $z^* \in Z$ and the value $\mu = \omega(z^*)$ for the player II is equivalent to solving the linear programming model as follows:

$$\min\{\omega\}$$
$$\text{s.t.}\begin{cases} \sum\limits_{j=1}^{n} a_{ij}z_j \le \omega \quad (i = 1, 2, \ldots, m) \\ \sum\limits_{j=1}^{n} z_j = 1 \\ z_j \ge 0 \quad (j = 1, 2, \ldots, n) \\ \omega \text{ unrestricted in sign.} \end{cases} \tag{1.8}$$

It is easy to see that Eqs. (1.5) and (1.8) are a pair of primal-dual linear programming models [1]. So the maximum of υ is equal to the minimum of ω. Their common value V is called the value of the matrix game $\boldsymbol{A}$, i.e., $V = \nu = \mu$.

1.3 Interval-Valued Mathematical Programming Models of Interval-Valued Matrix Games

1.3.1 Arithmetic Operations Over Intervals

Let R be the set of real numbers. An interval may be expressed as $\bar{a} = [a_L, a_R] = \{a | a_L \leq a \leq a_R, a_L \in \mathrm{R}, a_R \in \mathrm{R}\}$, where a_L and a_R are called the lower and upper bounds of the interval $\bar{a}$, respectively. The set of intervals in the real number set R is denoted by I(R).

If $a_L = a_R$, then the interval $\bar{a} = [a_L, a_R]$ degenerates to a real number a, where $a = a_L = a_R$. Conversely, a real number a can be written as an interval $\bar{a} = [a, a]$. Therefore, the concept of intervals is a generalization of that of real numbers. In other words, real numbers are special cases of intervals.

If $a_L \geq 0$, then $\bar{a} = [a_L, a_R]$ is called a non-negative interval, denoted by $\bar{a} \geq 0$. Likewise, if $a_R \leq 0$, then $\bar{a}$ is called a non-positive interval, denoted by $\bar{a} \leq 0$. If $a_L > 0$, then $\bar{a}$ is called a positive interval, denoted by $\bar{a} > 0$. If $a_R < 0$, then $\bar{a}$ is called a negative interval, denoted by $\bar{a} < 0$.

For any intervals $\bar{a} = [a_L, a_R]$ and $\bar{b} = [b_L, b_R]$, We stipulate their operations as follows:

(1) $\bar{a} + \bar{b} = [a_L, a_R] + [b_L, b_R] = [a_L + b_L, a_R + b_R]$;

(2) $\bar{a} - \bar{b} = [a_L, a_R] - [b_L, b_R] = [a_L - b_R, a_R - b_L]$;

(3) $\gamma\bar{a} = \gamma[a_L, a_R] = \begin{cases} [\gamma a_L, \gamma a_R] & \text{if } \gamma \geq 0 \\ [\gamma a_R, \gamma a_L] & \text{if } \gamma < 0 \end{cases}$;

(4) $\bar{a}\bar{b} = \bar{a} \times \bar{b} = [a_L, a_R] \times [b_L, b_R] = [\min\{a_L b_L, a_L b_R, a_R b_L, a_R b_R\}, \max\{a_L b_L, a_L b_R, a_R b_L, a_R b_R\}]$;

(5) $\frac{\bar{a}}{\bar{b}} = \frac{[a_L, a_R]}{[b_L, b_R]} = [\min\{\frac{a_L}{b_L}, \frac{a_L}{b_R}, \frac{a_R}{b_L}, \frac{a_R}{b_R}\}, \max\{\frac{a_L}{b_L}, \frac{a_L}{b_R}, \frac{a_R}{b_L}, \frac{a_R}{b_R}\}]$, provided that $b_L \leq b_R < 0$ or $b_R \geq b_L > 0$.

Alternatively, an interval $\bar{a}$ can be expressed in mean-width (or center-radius) form as $\bar{a} = \langle m(\bar{a}), w(\bar{a}) \rangle$, where $m(\bar{a}) = (a_L + a_R)/2$ and $w(\bar{a}) = (a_R - a_L)/2$ are the mid-point and half-width of the interval $\bar{a}$.

Using the aforementioned mean-width (or center-radius) form, we can rewrite the former three operations of the above intervals' operations as follows:

(1a) $\bar{a} + \bar{b} = \langle m(\bar{a}), w(\bar{a}) \rangle + \langle m(\bar{b}), w(\bar{b}) \rangle = \langle m(\bar{a}) + m(\bar{b}), w(\bar{a}) + w(\bar{b}) \rangle$;

(2a) $\bar{a} - \bar{b} = \langle m(\bar{a}), w(\bar{a}) \rangle - \langle m(\bar{b}), w(\bar{b}) \rangle = \langle m(\bar{a}) - m(\bar{b}), w(\bar{a}) + w(\bar{b}) \rangle$;

$$\text{(3a)}\quad \gamma\bar{a} = \gamma\langle m(\bar{a}), w(\bar{a})\rangle = \langle \gamma m(\bar{a}), |\gamma| w(\bar{a})\rangle = \begin{cases} \langle \gamma m(\bar{a}), \gamma w(\bar{a})\rangle & \text{if } \gamma \geq 0 \\ \langle \gamma m(\bar{a}), -\gamma w(\bar{a})\rangle & \text{if } \gamma < 0. \end{cases}$$

Example 1.1 Let $\bar{a}^0 = [1, 5]$ and $\bar{b}^0 = [2, 4]$ be two intervals, which represent the estimates of demand for some product. Thus, $\bar{a}^0 = [1, 5]$ and $\bar{b}^0 = [2, 4]$ can be rewritten as $\bar{a}^0 = \langle 3, 2\rangle$ and $\bar{b}^0 = \langle 3, 1\rangle$, respectively. Then, according to the above operations over intervals, we have:

$$\bar{a}^0 + \bar{b}^0 = [3, 9] = \langle 6, 3\rangle,$$
$$\bar{a}^0 - \bar{b}^0 = [-3, 3] = \langle 0, 3\rangle$$

and

$$3\bar{a}^0 = [3, 15] = \langle 9, 6\rangle.$$

Conversely, an interval $\bar{a} = \langle m(\bar{a}), w(\bar{a})\rangle$ in mean-width form can be easily rewritten as $\bar{a} = [m(\bar{a}) - w(\bar{a}), m(\bar{a}) + w(\bar{a})]$ in the ordinary form. For example, $\bar{a}^1 = \langle 3, 2\rangle$ is an interval in mean-width form. Thus, $\bar{a}^1$ can be rewritten as $\bar{a}^1 = [3 - 2, 3 + 2] = [1, 5]$, which is just about the interval $\bar{a}^0$ given in Example 1.1.

Let $\bar{a} = \langle m(\bar{a}), w(\bar{a})\rangle$ and $\bar{b} = \langle m(\bar{b}), w(\bar{b})\rangle$ be two intervals. For $m(\bar{a}) \leq m(\bar{b})$ and $w(\bar{a}) + w(\bar{b}) \neq 0$, an acceptability index to the premise $\bar{a} \leq_\psi \bar{b}$ is defined as follows [25]:

$$\psi(\bar{a} \leq_\psi \bar{b}) = \frac{m(\bar{b}) - m(\bar{a})}{w(\bar{b}) + w(\bar{a})}, \tag{1.9}$$

which is the value judgment or satisfaction degree of the player that the interval $\bar{a}$ is not superior to the interval $\bar{b}$ (or $\bar{b}$ is not inferior to $\bar{a}$) in terms of value. Here "not inferior to" and "not superior to" are analogous to "not less than" and "not greater than" in the real number set R, respectively. Similarly, the symbol "$\geq_\psi$" can be defined.

Thus, the orderings for two intervals $\bar{a}$ and $\bar{b}$ are defined as follows [25]:

$$\bar{a} \vee_\psi \bar{b} = \begin{cases} \bar{b} & \text{if } \psi(\bar{a} \leq_\psi \bar{b}) > 0 \\ \bar{a} & \text{if } \psi(\bar{a} \leq_\psi \bar{b}) = 0 \text{ and } w(\bar{a}) < w(\bar{b}) \text{ and player is pessimistic} \\ \bar{b} & \text{if } \psi(\bar{a} \leq_\psi \bar{b}) = 0 \text{ and } w(\bar{a}) < w(\bar{b}) \text{ and player is optimistic} \end{cases} \tag{1.10}$$

and

$$\bar{a} \wedge_\psi \bar{b} = \begin{cases} \bar{b} & \text{if } \psi(\bar{b} \leq_\psi \bar{a}) > 0 \\ \bar{a} & \text{if } \psi(\bar{b} \leq_\psi \bar{a}) = 0 \text{ and } w(\bar{a}) > w(\bar{b}) \text{ and player is pessimistic} \\ \bar{b} & \text{if } \psi(\bar{b} \leq_\psi \bar{a}) = 0 \text{ and } w(\bar{a}) > w(\bar{b}) \text{ and player is optimistic.} \end{cases} \tag{1.11}$$

In the sequent discussions, the max operator "$\vee_{\psi}$" in Eq. (1.10) and the min operator "$\wedge_{\psi}$" in Eq. (1.11) are meant to be in the sense of Eq. (1.9) unless specially stated. Sometimes, $\bar{a} \vee_{\psi} \bar{b}$ and $\bar{a} \wedge_{\psi} \bar{b}$ may be written as $\max_{\psi}\{\bar{a}, \bar{b}\}$ and $\min_{\psi}\{\bar{a}, \bar{b}\}$, respectively.

1.3.2 *Concepts of Solutions of Interval-Valued Matrix Games and Properties*

As stated in Sect. 1.2, the sets of pure strategies for the players I and II are $S_1 = \{\delta_1, \delta_2, \ldots, \delta_m\}$ and $S_2 = \{\beta_1, \beta_2, \ldots, \beta_n\}$; the sets of mixed strategies for the players I and II are Y and Z, respectively. If the player I adopts any pure strategy $\delta_i \in S_1$ and the player II adopts any pure strategy $\beta_j \in S_2$, then the payoff of the player I is expressed with an interval $\bar{a}_{ij} = [a_{Lij}, a_{Rij}]$ and hereby the payoff of the player II is equal to $-\bar{a}_{ij} = [-a_{Rij}, -a_{Lij}]$. The interval-valued payoff matrix for the player I can be concisely expressed in the matrix form as follows:

$$\bar{\boldsymbol{A}} = (\bar{a}_{ij})_{m\times n} = \begin{matrix} \\ \delta_1 \\ \delta_2 \\ \vdots \\ \delta_m \end{matrix} \begin{matrix} \begin{matrix} \beta_1 & \beta_2 & \cdots & \beta_n \end{matrix} \\ \begin{pmatrix} [a_{L11}, a_{R11}] & [a_{L12}, a_{R12}] & \cdots & [a_{L1n}, a_{R1n}] \\ [a_{L21}, a_{R21}] & [a_{L22}, a_{R22}] & \cdots & [a_{L2n}, a_{R2n}] \\ \vdots & \vdots & \cdots & \vdots \\ [a_{Lm1}, a_{Rm1}] & [a_{Lm2}, a_{Rm2}] & \cdots & [a_{Lmn}, a_{Rmn}] \end{pmatrix} \end{matrix}$$

and hereby the interval-valued payoff matrix for the player II is equal to

$$-\bar{\boldsymbol{A}} = (-\bar{a}_{ij})_{m\times n} = \begin{matrix} \\ \delta_1 \\ \delta_2 \\ \vdots \\ \delta_m \end{matrix} \begin{matrix} \begin{matrix} \beta_1 & \beta_2 & \cdots & \beta_n \end{matrix} \\ \begin{pmatrix} [-a_{R11}, -a_{L11}] & [-a_{R12}, -a_{L12}] & \cdots & [-a_{R1n}, -a_{L1n}] \\ [-a_{R21}, -a_{L21}] & [-a_{R22}, -a_{L22}] & \cdots & [-a_{R2n}, -a_{L2n}] \\ \vdots & \vdots & \cdots & \vdots \\ [-a_{Rm1}, -a_{Lm1}] & [-a_{Rm2}, -a_{Lm2}] & \cdots & [-a_{Rmn}, -a_{Lmn}] \end{pmatrix} \end{matrix}.$$

In the sequent, an interval-valued matrix game with the interval-valued payoff matrix $\bar{\boldsymbol{A}}$ is called as the interval-valued matrix game $\bar{\boldsymbol{A}}$ for short.

In a similar way to the classical matrix game [1, 25], if both $\underset{1\le i\le m}{\vee_{\psi}} \{ \underset{1\le j\le n}{\wedge_{\psi}} \{[a_{Lij}, a_{Rij}]\}\}$ and $\underset{1\le j\le n}{\wedge_{\psi}} \{ \underset{1\le i\le m}{\vee_{\psi}} \{[a_{Lij}, a_{Rij}]\}\}$ exist and are identical, i.e.,

$$[a_{Lkr}, a_{Rkr}] = \underset{1\le i\le m}{\vee_{\psi}} \{ \underset{1\le j\le n}{\wedge_{\psi}} \{[a_{Lij}, a_{Rij}]\}\} = \underset{1\le j\le n}{\wedge_{\psi}} \{ \underset{1\le i\le m}{\vee_{\psi}} \{[a_{Lij}, a_{Rij}]\}\},$$

then, the interval-valued matrix game $\bar{\boldsymbol{A}}$ has a saddle point (*k*, *r*) or (δ_k, β_r) (in the sense of pure strategies). An interval-valued matrix game $\bar{\boldsymbol{A}}$ is strictly determined if it has a saddle point. However, it is not always true that any interval-valued matrix game is strictly determined. The reason is that some interval-valued matrix games may have no saddle points in the sense of the pure strategies even if both $\underset{1\le i\le m}{\vee_\psi}\{\underset{1\le j\le n}{\wedge_\psi}\{[a_{Lij}, a_{Rij}]\}\}$ and $\underset{1\le j\le n}{\wedge_\psi}\{\underset{1\le i\le m}{\vee_\psi}\{[a_{Lij}, a_{Rij}]\}\}$ exist whereas they may not be always identical.

Example 1.2 Let us consider the 2 × 2 interval-valued matrix game $\bar{\boldsymbol{A}}_1$ as follows [25, 26]:

$$\bar{\boldsymbol{A}}_1 = ([a_{Lij}, a_{Rij}])_{2\times 2} = \begin{matrix} \\ \alpha_1 \\ \alpha_2 \end{matrix} \begin{matrix} \begin{matrix} \beta_1 & \quad \beta_2 \end{matrix} \\ \begin{pmatrix} [-1, 1] & [1, 3] \\ [0, 2] & [-2, 0] \end{pmatrix} \end{matrix}.$$

Obviously, we have

$$\underset{1\le i\le 2}{\vee_\psi}\{\underset{1\le j\le 2}{\wedge_\psi}\{[a_{Lij}, a_{Rij}]\}\} = [a_{L11}, a_{R11}] = [-1, 1].$$

and

$$\underset{1\le j\le 2}{\wedge_\psi}\{\underset{1\le i\le 2}{\vee_\psi}\{[a_{Lij}, a_{Rij}]\}\} = [a_{L21}, a_{R21}] = [0, 2].$$

Hence, we obtain

$$\underset{1\le i\le 2}{\vee_\psi}\{\underset{1\le j\le 2}{\wedge_\psi}\{[a_{Lij}, a_{Rij}]\}\} \neq \underset{1\le j\le 2}{\wedge_\psi}\{\underset{1\le i\le 2}{\vee_\psi}\{[a_{Lij}, a_{Rij}]\}\}.$$

Therefore, the interval-valued matrix game $\bar{\boldsymbol{A}}_1$ has no saddle points. Thus, the interval-valued matrix game $\bar{\boldsymbol{A}}_1$ is not strictly determined. Thus, we need to introduce the concept of solutions of interval-valued matrix games in the sense of mixed strategies.

If the player I chooses a mixed strategy $\boldsymbol{y} \in Y$ and II chooses a mixed strategy $\boldsymbol{z} \in Z$, then according to the arithmetic operations over intervals stated as in Sect. 1.3.1, the interval-valued expected payoff of the player I can be computed as follows:

$$\boldsymbol{y}^{\mathrm{T}}\bar{\boldsymbol{A}}\boldsymbol{z} = \sum_{i=1}^{m}\sum_{j=1}^{n} y_i[a_{Lij}, a_{Rij}]z_j = [\sum_{i=1}^{m}\sum_{j=1}^{n} y_i a_{Lij} z_j, \sum_{i=1}^{m}\sum_{j=1}^{n} y_i a_{Rij} z_j].$$

Due to the fact that the interval-valued matrix game $\bar{A}$ is zero-sum, therefore the interval-valued expected payoff of the player II is equal to

$$y^{\mathrm{T}}(-\bar{A})z = \sum_{i=1}^{m}\sum_{j=1}^{n} y_i[-a_{Rij}, -a_{Lij}]z_j = [-\sum_{i=1}^{m}\sum_{j=1}^{n} y_i a_{Rij} z_j, -\sum_{i=1}^{m}\sum_{j=1}^{n} y_i a_{Lij} z_j].$$

Definition 1.1 Let $\bar{\upsilon} = [\upsilon_L, \upsilon_R]$ and $\bar{\omega} = [\omega_L, \omega_R]$ be two intervals on the real number set R. Assume that there exist mixed strategies $\boldsymbol{y}^* \in Y$ and $\boldsymbol{z}^* \in Z$. If for any mixed strategies $\boldsymbol{y} \in Y$ and $\boldsymbol{z} \in Z$, $(\boldsymbol{y}^*, \boldsymbol{z}^*, \bar{\upsilon}, \bar{\omega})$ satisfies both $\boldsymbol{y}^{*\mathrm{T}}\bar{\boldsymbol{A}}\boldsymbol{z} \geq_{\psi} \bar{\upsilon}$ and $\boldsymbol{y}^{\mathrm{T}}\bar{\boldsymbol{A}}\boldsymbol{z}^* \leq_{\psi} \bar{\omega}$, then $(\boldsymbol{y}^*, \boldsymbol{z}^*, \bar{\upsilon}, \bar{\omega})$ is called a reasonable solution of the interval-valued matrix game $\bar{\boldsymbol{A}}$; $\bar{\upsilon}$ and $\bar{\omega}$ are called reasonable values for the players I and II, respectively; $\boldsymbol{y}^*$ and $\boldsymbol{z}^*$ are called reasonable strategies for players I and II, respectively.

Let U and W be the sets of reasonable values $\bar{\upsilon}$ and $\bar{\omega}$ for players I and II, respectively.

It is worth noticing that Definition 1.1 only gives the notion of reasonable solutions of interval-valued matrix games rather than the notion of optimal solutions. In other words, the reasonable solution is not the solution of the interval-valued matrix game, which is given as follows.

Definition 1.2 Assume that there exist two reasonable values $\bar{\upsilon}^* \in U$ and $\bar{\omega}^* \in W$. If there do not exist reasonable values $\bar{\upsilon}' \in U$ $(\bar{\upsilon}' \neq \bar{\upsilon}^*)$ and $\bar{\omega}' \in W$ $(\bar{\omega}' \neq \bar{\omega}^*)$ such that they satisfy both $\bar{\upsilon}' \geq_{\psi} \bar{\upsilon}^*$ and $\bar{\omega}' \leq_{\psi} \bar{\omega}^*$, then $(\boldsymbol{y}^*, \boldsymbol{z}^*, \bar{\upsilon}^*, \bar{\omega}^*)$ is called a solution of the interval-valued matrix game $\bar{\boldsymbol{A}}$; $\boldsymbol{y}^*$ is called an optimal (or a maximin) strategy for the player I and $\boldsymbol{z}^*$ is called an optimal (or a minimax) strategy for the player II; $\bar{\upsilon}^*$ and $\bar{\omega}^*$ are called the player I's gain-floor and the player II's loss-ceiling, respectively.

Thus, we draw an important and a useful conclusion, which can be summarized as in Theorem 1.1.

Theorem 1.1 *For any interval-valued matrix game* $\bar{\boldsymbol{A}}$, *both* $\bigvee_{\psi\, \boldsymbol{y}\in Y}\{\bigwedge_{\psi\, \boldsymbol{z}\in Z}\{\boldsymbol{y}^{\mathrm{T}}\bar{\boldsymbol{A}}\boldsymbol{z}\}\}$ *and* $\bigwedge_{\psi\, \boldsymbol{z}\in Z}\{\bigvee_{\psi\, \boldsymbol{y}\in Y}\{\boldsymbol{y}^{\mathrm{T}}\bar{\boldsymbol{A}}\boldsymbol{z}\}\}$ *exist and the following inequality is valid*

$$\bigvee_{\psi\, \boldsymbol{y}\in Y}\{\bigwedge_{\psi\, \boldsymbol{z}\in Z}\{\boldsymbol{y}^{\mathrm{T}}\bar{\boldsymbol{A}}\boldsymbol{z}\}\} \leq_{\psi} \bigwedge_{\psi\, \boldsymbol{z}\in Z}\{\bigvee_{\psi\, \boldsymbol{y}\in Y}\{\boldsymbol{y}^{\mathrm{T}}\bar{\boldsymbol{A}}\boldsymbol{z}\}\}.$$

Proof Obviously, for any strategies $\boldsymbol{y} \in Y$ and $\boldsymbol{z} \in Z$, we have

$$\bigwedge_{\psi\, \boldsymbol{z}\in Z}\{\boldsymbol{y}^{\mathrm{T}}\bar{\boldsymbol{A}}\boldsymbol{z}\} \leq_{\psi} \boldsymbol{y}^{\mathrm{T}}\bar{\boldsymbol{A}}\boldsymbol{z} \leq_{\psi} \bigvee_{\psi\, \boldsymbol{y}\in Y}\{\boldsymbol{y}^{\mathrm{T}}\bar{\boldsymbol{A}}\boldsymbol{z}\}$$

Hence, it follows that

$$\bigvee_{\psi\, \boldsymbol{y}\in Y}\{\bigwedge_{\psi\, \boldsymbol{z}\in Z}\{\boldsymbol{y}^{\mathrm{T}}\bar{\boldsymbol{A}}\boldsymbol{z}\}\} \leq_{\psi} \bigvee_{\psi\, \boldsymbol{y}\in Y}\{\boldsymbol{y}^{\mathrm{T}}\bar{\boldsymbol{A}}\boldsymbol{z}\}.$$

Therefore, we obtain

$$\vee_{\psi}_{y\in Y}\{\wedge_{\psi}_{z\in Z}\{\mathbf{y}^{\mathrm{T}}\bar{\mathbf{A}}\mathbf{z}\}\} \leq_{\psi} \wedge_{\psi}_{z\in Z}\{\vee_{\psi}_{y\in Y}\{\mathbf{y}^{\mathrm{T}}\bar{\mathbf{A}}\mathbf{z}\}\}.$$

Thus, the proof of Theorem 1.1 has been completed.

It is easy to see from Definitions 1.1 and 1.2 that $\bar{\upsilon}^* = \vee_{\psi}_{y\in Y}\{\wedge_{\psi}_{z\in Z}\{\mathbf{y}^{\mathrm{T}}\bar{\mathbf{A}}\mathbf{z}\}\}$ and $\bar{\omega}^* = \wedge_{\psi}_{z\in Z}\{\vee_{\psi}_{y\in Y}\{\mathbf{y}^{\mathrm{T}}\bar{\mathbf{A}}\mathbf{z}\}\}$. Then, we have

$$\bar{\upsilon}^* \leq_{\psi} \bar{\omega}^*. \tag{1.12}$$

Thus, Theorem 1.1 means that the player I's gain-floor "essentially cannot exceed" the player II's loss-ceiling.

It is worthwhile to note that Eq. (1.12) is strictly valid for some interval-valued matrix games, i.e., $\bar{\upsilon}^* <_{\psi} \bar{\omega}^*$. For example, it is easy to see from the 2 × 2 interval-valued matrix game $\bar{\mathbf{A}}_1$ given in Example 1.2 that

$$\vee_{\psi}_{1\leq i\leq 2}\{\wedge_{\psi}_{1\leq j\leq 2}\{[a_{Lij}, a_{Rij}]\}\} = [-1, 1] <_{\psi} [0, 2] = \wedge_{\psi}_{1\leq j\leq 2}\{\vee_{\psi}_{1\leq i\leq 2}\{[a_{Lij}, a_{Rij}]\}\}.$$

Note that pure strategies are special cases of mixed strategies and the payoffs are intervals. In other words, there may not always exist an interval $\bar{v} \in I(\mathrm{R})$ such that it satisfies both $\mathbf{y}^{*\mathrm{T}}\bar{\mathbf{A}}\mathbf{z} \geq_{\psi} \bar{v}$ and $\mathbf{y}^{\mathrm{T}}\bar{\mathbf{A}}\mathbf{z}^* \leq_{\psi} \bar{v}$ for any mixed strategies $\mathbf{y} \in Y$ and $\mathbf{z} \in Z$.

1.3.3 Auxiliary Interval-Valued Mathematical Programming Models

It is easy to see that the strategy spaces Y and Z of the players I and II are finite, compact, and convex sets. Hence, $\wedge_{\psi}_{z\in Z}\{\mathbf{y}^{\mathrm{T}}\bar{\mathbf{A}}\mathbf{z}\}$ and $\vee_{\psi}_{y\in Y}\{\mathbf{y}^{\mathrm{T}}\bar{\mathbf{A}}\mathbf{z}\}$ will be attained at the extreme points of the strategy spaces Y and Z, respectively. Thus, for any mixed strategies $\mathbf{y} \in Y$ and $\mathbf{z} \in Z$, we have

$$\vee_{\psi}_{y\in Y}\{\wedge_{\psi}_{z\in Z}\{\mathbf{y}^{\mathrm{T}}\bar{\mathbf{A}}\mathbf{z}\}\} = \vee_{\psi}_{y\in Y}\{\wedge_{\psi}_{1\leq j\leq n}\{\sum_{i=1}^{m}[a_{Lij}, a_{Rij}]y_i\}\}$$

and

$$\wedge_{\psi}_{z\in Z}\{\vee_{\psi}_{y\in Y}\{\mathbf{y}^{\mathrm{T}}\bar{\mathbf{A}}\mathbf{z}\}\} = \wedge_{\psi}_{z\in Z}\{\vee_{\psi}_{1\leq i\leq m}\{\sum_{j=1}^{m}[a_{Lij}, a_{Rij}]z_i\}\},$$

respectively.

Denote

$$\bar{\upsilon} = [\upsilon_L, \upsilon_R] = \wedge_{\psi}{}_{1 \le j \le n} \{\sum_{i=1}^{m} [a_{Lij}, a_{Rij}] y_i\}$$

and

$$\bar{\omega} = [\omega_L, \omega_R] = \vee_{\psi}{}_{1 \le i \le m} \{\sum_{j=1}^{n} [a_{Lij}, a_{Rij}] z_j\}.$$

Then, the maximin value of $\bar{\upsilon}$ for the player I and the minimax value of $\bar{\omega}$ for the player II can be obtained through solving the interval-valued mathematical programming models as follows:

$$\max_{\psi}\{\bar{\upsilon}\}$$
$$\text{s.t.}\begin{cases} \sum_{i=1}^{m} [a_{Lij}, a_{Rij}] y_i \ge_{\psi} \bar{\upsilon} \quad (j = 1, 2, \ldots, n) \\ \sum_{i=1}^{m} y_i = 1 \\ y_i \ge 0 \quad (i = 1, 2, \ldots, m) \\ \bar{\upsilon} \text{ unrestricted in sign} \end{cases} \tag{1.13}$$

and

$$\min_{\psi}\{\bar{\omega}\}$$
$$\text{s.t.}\begin{cases} \sum_{j=1}^{n} [a_{Lij}, a_{Rij}] z_j \le_{\psi} \bar{\omega} \quad (i = 1, 2, \ldots, m) \\ \sum_{j=1}^{n} z_j = 1 \\ z_j \ge 0 \quad (j = 1, 2, \ldots, n) \\ \bar{\omega} \text{ unrestricted in sign,} \end{cases} \tag{1.14}$$

respectively.

Assume that there are two interval-valued matrix games $\bar{\boldsymbol{A}} = ([a_{Lij}, a_{Rij}])_{m \times n}$ and $\bar{\boldsymbol{A}}'' = ([a''_{Lij}, a''_{Rij}])_{m \times n}$, where $[c_L, c_R]$ is an interval and $[a''_{Lij}, a''_{Rij}] = [c_L, c_R] + [a_{Lij}, a_{Rij}] = [c_L + a_{Lij}, c_R + a_{Rij}]$. Denote $\bar{\boldsymbol{A}}'' = ([c_L, c_R])_{m \times n} + \bar{\boldsymbol{A}}$, i.e.,

$$\begin{array}{c} \\ \delta_1 \\ \delta_2 \\ \vdots \\ \delta_m \end{array}\begin{array}{c} \begin{array}{cccc} \beta_1 & \beta_2 & \cdots & \beta_n \end{array} \\ \begin{pmatrix} [a''_{L11}, a''_{R11}] & [a''_{L12}, a''_{R12}] & \cdots & [a''_{L1n}, a''_{R1n}] \\ [a''_{L21}, a''_{R21}] & [a''_{L22}, a''_{R22}] & \cdots & [a''_{L2n}, a''_{R2n}] \\ \vdots & \vdots & \cdots & \vdots \\ [a''_{Lm1}, a''_{Rm1}] & [a''_{Lm2}, a''_{Rm2}] & \cdots & [a''_{Lmn}, a''_{Rmn}] \end{pmatrix} \end{array}$$

$$= \begin{array}{c} \begin{array}{cccc} \beta_1 & \beta_2 & \cdots & \beta_n \end{array} \\ \begin{pmatrix} [c_L, c_R] & [c_L, c_R] & \cdots & [c_L, c_R] \\ [c_L, c_R] & [c_L, c_R] & \cdots & [c_L, c_R] \\ \vdots & \vdots & \cdots & \vdots \\ [c_L, c_R] & [c_L, c_R] & \cdots & [c_L, c_R] \end{pmatrix} \end{array}$$

$$+ \begin{pmatrix} [a_{L11}, a_{R11}] & [a_{L12}, a_{R12}] & \cdots & [a_{L1n}, a_{R1n}] \\ [a_{L21}, a_{R21}] & [a_{L22}, a_{R22}] & \cdots & [a_{L2n}, a_{R2n}] \\ \vdots & \vdots & \cdots & \vdots \\ [a_{Lm1}, a_{Rm1}] & [a_{Lm2}, a_{Rm2}] & \cdots & [a_{Lmn}, a_{Rmn}] \end{pmatrix}$$

$$= \begin{array}{c} \\ \delta_1 \\ \delta_2 \\ \vdots \\ \delta_m \end{array}\begin{array}{c} \begin{array}{cccc} \beta_1 & \beta_2 & \cdots & \beta_n \end{array} \\ \begin{pmatrix} [c_L + a_{L11}, c_R + a_{R11}] & [c_L + a_{L12}, c_R + a_{R12}] & \cdots & [c_L + a_{L1n}, c_R + a_{R1n}] \\ [c_L + a_{L21}, c_R + a_{R21}] & [c_L + a_{L22}, c_R + a_{R22}] & \cdots & [c_L + a_{L2n}, c_R + a_{R2n}] \\ \vdots & \vdots & \cdots & \vdots \\ [c_L + a_{Lm1}, c_R + a_{Rm1}] & [c_L + a_{Lm2}, c_R + a_{Rm2}] & \cdots & [c_L + a_{Lmn}, c_R + a_{Rmn}] \end{pmatrix} \end{array}.$$

Thus, we can draw an important property on the relation between the interval-valued matrix games $\bar{\boldsymbol{A}}$ and $\bar{\boldsymbol{A}}''$, which is summarized as in Theorem 1.2.

Theorem 1.2 *Assume that* $\bar{\boldsymbol{A}}'' = ([c_L, c_R])_{m \times n} + \bar{\boldsymbol{A}}$. *Then, the interval-valued matrix games* $\bar{\boldsymbol{A}}$ *and* $\bar{\boldsymbol{A}}''$ *have the identical optimal strategies for the players I and II, and* $\bar{\upsilon}''^* = [c_L, c_R] + \bar{\upsilon}^*$, $\bar{\omega}''^* = [c_L, c_R] + \bar{\omega}^*$, *where* $\bar{\upsilon}^*$ *and* $\bar{\upsilon}''^*$ *are the player I's gain-floors in the interval-valued matrix games* $\bar{\boldsymbol{A}}$ *and* $\bar{\boldsymbol{A}}''$, $\bar{\omega}^*$ *and* $\bar{\omega}''^*$ *are the player II's loss-ceilings in* $\bar{\boldsymbol{A}}$ *and* $\bar{\boldsymbol{A}}''$, *respectively.*

Proof For any interval $[c_L, c_R]$, it is easily derived from Eqs. (1.13) and (1.14) that

$$\max_{\psi}\{[c_L, c_R] + \bar{\upsilon}\}$$

$$\text{s.t.} \begin{cases} \sum_{i=1}^{m} ([c_L, c_R] + [a_{Lij}, a_{Rij}]) y_i \geq_{\psi} [c_L, c_R] + \bar{\upsilon} \quad (j = 1, 2, \ldots, n) \\ \sum_{i=1}^{m} y_i = 1 \\ y_i \geq 0 \quad (i = 1, 2, \ldots, m) \\ \bar{\upsilon} \text{ unrestricted in sign} \end{cases}$$

and

$$\min_{\psi}\{[c_L, c_R] + \bar{\omega}\}$$

$$\text{s.t.}\begin{cases} \sum_{j=1}^{n}([c_L, c_R] + [a_{Lij}, a_{Rij}])z_j \leq_{\psi} [c_L, c_R] + \bar{\omega}\,(i = 1, 2, \cdots, m) \\ \sum_{j=1}^{n} z_j = 1 \\ z_j \geq 0 \quad (j = 1, 2, \ldots, n) \\ \bar{\omega} \text{ unrestricted in sign,} \end{cases}$$

respectively. Namely, we have

$$\max_{\psi}\{[c_L, c_R] + \bar{\upsilon}\}$$

$$\text{s.t.}\begin{cases} \sum_{i=1}^{m}([c_L + a_{Lij}, c_R + a_{Rij}])y_i \geq_{\psi} [c_L, c_R] + \bar{\upsilon} \quad (j = 1, 2, \ldots, n) \\ \sum_{i=1}^{m} y_i = 1 \\ y_i \geq 0 \quad (i = 1, 2, \ldots, m) \\ \bar{\upsilon} \text{ unrestricted in sign} \end{cases}$$

and

$$\min_{\psi}\{[c_L, c_R] + \bar{\omega}\}$$

$$\text{s.t.}\begin{cases} \sum_{j=1}^{n}([c_L + a_{Lij}, c_R + a_{Rij}])z_j \leq_{\psi} [c_L, c_R] + \bar{\omega} \quad (i = 1, 2, \ldots, m) \\ \sum_{j=1}^{n} z_j = 1 \\ z_j \geq 0 \quad (j = 1, 2, \ldots, n) \\ \bar{\omega} \text{ unrestricted in sign,} \end{cases}$$

which are just about a pair of interval-valued mathematical programming models for the interval-valued matrix game $\bar{\boldsymbol{A}}^{''}$. Thus, the proof of Theorem 1.2 has been completed.

For any positive interval $[c_L, c_R]$, i.e., $c_L > 0$, we denote an interval-valued payoff matrix by $\bar{\boldsymbol{A}}^{'} = ([a^{'}_{Lij}, a^{'}_{Rij}])_{m\times n}$, where $[a^{'}_{Lij}, a^{'}_{Rij}] = [c_L, c_R] \times [a_{Lij}, a_{Rij}] = [c_L a_{Lij}, c_R a_{Rij}]$. Denote $\bar{\boldsymbol{A}}^{'} = [c_L, c_R] \times \bar{\boldsymbol{A}}$ or $\bar{\boldsymbol{A}}^{'} = [c_L, c_R]\bar{\boldsymbol{A}}$, i.e.,

$$
\begin{array}{c} \\ \delta_1 \\ \delta_2 \\ \vdots \\ \delta_m \end{array}
\begin{array}{c}
\begin{array}{cccc} \beta_1 & \beta_2 & \cdots & \beta_n \end{array} \\
\begin{pmatrix}
[a'_{L11}, a'_{R11}] & [a'_{L12}, a'_{R12}] & \cdots & [a'_{L1n}, a'_{R1n}] \\
[a'_{L21}, a'_{R21}] & [a'_{L22}, a'_{R22}] & \cdots & [a'_{L2n}, a'_{R2n}] \\
\vdots & \vdots & \cdots & \vdots \\
[a'_{Lm1}, a'_{Rm1}] & [a'_{Lm2}, a'_{Rm2}] & \cdots & [a'_{Lmn}, a'_{Rmn}]
\end{pmatrix}
\end{array}
$$

$$
= [c_L, c_R] \times
\begin{array}{c}
\begin{array}{cccc} \beta_1 & \beta_2 & \cdots & \beta_n \end{array} \\
\begin{pmatrix}
[a_{L11}, a_{R11}] & [a_{L12}, a_{R12}] & \cdots & [a_{L1n}, a_{R1n}] \\
[a_{L21}, a_{R21}] & [a_{L22}, a_{R22}] & \cdots & [a_{L2n}, a_{R2n}] \\
\vdots & \vdots & \cdots & \vdots \\
[a_{Lm1}, a_{Rm1}] & [a_{Lm2}, a_{Rm2}] & \cdots & [a_{Lmn}, a_{Rmn}]
\end{pmatrix}
\end{array}
$$

$$
= \begin{array}{c} \\ \delta_1 \\ \delta_2 \\ \vdots \\ \delta_m \end{array}
\begin{array}{c}
\begin{array}{cccc} \beta_1 & \beta_2 & \cdots & \beta_n \end{array} \\
\begin{pmatrix}
[c_L a_{L11}, c_R a_{R11}] & [c_L a_{L12}, c_R a_{R12}] & \cdots & [c_L a_{L1n}, c_R a_{R1n}] \\
[c_L a_{L21}, c_R a_{R21}] & [c_L a_{L22}, c_R a_{R22}] & \cdots & [c_L a_{L2n}, c_R a_{R2n}] \\
\vdots & \vdots & \cdots & \vdots \\
[c_L a_{Lm1}, c_R a_{Rm1}] & [c_L a_{Lm2}, c_R a_{Rm2}] & \cdots & [c_L a_{Lmn}, c_R a_{Rmn}]
\end{pmatrix}
\end{array}.
$$

In a similar way to Theorem 1.2, we draw the following important property.

Theorem 1.3 *Assume that* $\bar{\boldsymbol{A}}' = [c_L, c_R]\bar{\boldsymbol{A}}$. *Then, the interval-valued matrix games* $\bar{\boldsymbol{A}}$ *and* $\bar{\boldsymbol{A}}'$ *have the identical optimal strategies for the players I and II, and* $\bar{\upsilon}'^* = [c_L, c_R]\bar{\upsilon}^*$, $\bar{\omega}'^* = [c_L, c_R]\bar{\omega}^*$, *where* $\bar{\upsilon}^*$ *and* $\bar{\upsilon}'^*$ *are the player I's gain-floors in the interval-valued matrix games* $\bar{\boldsymbol{A}}$ *and* $\bar{\boldsymbol{A}}'$, $\bar{\omega}^*$ *and* $\bar{\omega}'^*$ *are the player II's loss-ceilings in* $\bar{\boldsymbol{A}}$ *and* $\bar{\boldsymbol{A}}'$, *respectively.*

Proof Since $[c_L, c_R]$ is a positive interval, it is easily derived from Eqs. (1.13) and (1.14) that

$$
\max_{\psi}\{[c_L, c_R]\bar{\upsilon}\}
$$

$$
\text{s.t.}\begin{cases}
\sum_{i=1}^{m} [c_L, c_R][a_{Lij}, a_{Rij}]y_i \geq_{\psi} [c_L, c_R]\bar{\upsilon} \quad (j = 1, 2, \ldots, n) \\
\sum_{i=1}^{m} y_i = 1 \\
y_i \geq 0 \quad (i = 1, 2, \ldots, m) \\
\bar{\upsilon} \text{ unrestricted in sign}
\end{cases}
$$

and

$$\min_{\psi}\{[c_L, c_R]\bar{\omega}\}$$

$$\text{s.t.}\begin{cases} \sum_{j=1}^{n}[c_L, c_R][a_{Lij}, a_{Rij}]z_j \leq_{\psi} [c_L, c_R]\bar{\omega} \quad (i = 1, 2, \ldots, m) \\ \sum_{j=1}^{n} z_j = 1 \\ z_j \geq 0 \quad (j = 1, 2, \ldots, n) \\ \bar{\omega} \text{ unrestricted in sign,} \end{cases}$$

respectively. Namely, we have

$$\max_{\psi}\{[c_L, c_R]\bar{\upsilon}\}$$

$$\text{s.t.}\begin{cases} \sum_{i=1}^{m}[c_L a_{Lij}, c_R a_{Rij}]y_i \geq_{\psi} [c_L, c_R]\bar{\upsilon} \quad (j = 1, 2, \ldots, n) \\ \sum_{i=1}^{m} y_i = 1 \\ y_i \geq 0 \quad (i = 1, 2, \ldots, m) \\ \bar{\upsilon} \text{ unrestricted in sign} \end{cases}$$

and

$$\min_{\psi}\{[c_L, c_R]\bar{\omega}\}$$

$$\text{s.t.}\begin{cases} \sum_{j=1}^{n}[c_L a_{Lij}, c_R a_{Rij}]z_j \leq_{\psi} [c_L, c_R]\bar{\omega} \quad (i = 1, 2, \ldots, m) \\ \sum_{j=1}^{n} z_j = 1 \\ z_j \geq 0 \quad (j = 1, 2, \ldots, n) \\ \bar{\omega} \text{ unrestricted in sign,} \end{cases}$$

which are just about a pair of interval-valued mathematical programming models for the interval-valued matrix game $\bar{A}'$. Thus, the proof of Theorem 1.3 has been completed.

1.3.4 Solving Methods of 2 × 2 Interval-Valued Matrix Games

Assume that the interval-valued payoff matrix of any 2 × 2 interval-valued matrix game is given as follows:

$$\bar{A} = ([a_{Lij}, a_{Rij}])_{2\times 2} = \begin{matrix} & \beta_1 & \beta_2 \\ \delta_1 & [a_{L11}, a_{R11}] & [a_{L12}, a_{R12}] \\ \delta_2 & [a_{L21}, a_{R21}] & [a_{R22}, a_{R22}] \end{matrix} .$$

Corollary 1.1 *For the* 2×2 *interval-valued matrix game* $\bar{A} = ([a_{Lij}, a_{Rij}])_{2\times 2}$, *if its intervals* $[a_{Lij}, a_{Rij}]$ *satisfy the conditions:* $a_{Rij} = a_{Lij} + \mu$ $(i, j = 1, 2)$, *where* $\mu > 0$, *then the optimal strategies for the players I and II and the value of the* 2×2 *interval-valued matrix game* $\bar{A}$ *are obtained as follows:*

$$y^* = (y_1^*, y_2^*)^{\mathrm{T}} = \left(\frac{a_{L22} - a_{L21}}{a_{L11} + a_{L22} - a_{L12} - a_{L21}}, \frac{a_{L11} - a_{L12}}{a_{L11} + a_{L22} - a_{L12} - a_{L21}}\right)^{\mathrm{T}}, \quad (1.15)$$

$$z^* = (z_1^*, z_2^*)^{\mathrm{T}} = \left(\frac{a_{L22} - a_{L12}}{a_{L11} + a_{L22} - a_{L12} - a_{L21}}, \frac{a_{L11} - a_{L21}}{a_{L11} + a_{L22} - a_{L12} - a_{L21}}\right)^{\mathrm{T}} \quad (1.16)$$

and

$$\bar{v}^* = [v_L^*, v_R^*] = \left[\frac{a_{L11}a_{L22} - a_{12}a_{21}}{a_{L11} + a_{L22} - a_{L12} - a_{L21}}, \frac{a_{R11}a_{R22} - a_{R12}a_{R21}}{a_{R11} + a_{R22} - a_{R12} - a_{R21}}\right], \quad (1.17)$$

respectively.

Proof It directly follows from $a_{Rij} = a_{Lij} + \mu$ $(i, j = 1, 2)$ that

$$[a'_{Lij}, a'_{Rij}] = \frac{[a_{Lij}, a_{Rij}]}{\mu} = \left[\frac{a_{Lij}}{\mu}, \frac{a_{Lij} + \mu}{\mu}\right] = \left[\frac{a_{Lij}}{\mu}, \frac{a_{Lij}}{\mu} + 1\right],$$

i.e., $a'_{Rij} = a'_{Lij} + 1$ $(i, j = 1, 2)$. Hence, it is easy to verify that the two following equations are valid:

$$\frac{a'_{L22} - a'_{L21}}{a'_{L11} + a'_{L22} - a'_{L12} - a'_{L21}} = \frac{a'_{R22} - a'_{R21}}{a'_{R11} + a'_{R22} - a'_{R12} - a'_{R21}}$$

and

$$\frac{a'_{L22} - a'_{L12}}{a'_{L11} + a'_{L22} - a'_{L12} - a'_{L21}} = \frac{a'_{R22} - a'_{R12}}{a'_{R11} + a'_{R22} - a'_{R12} - a'_{R21}}.$$

Then, using the similar method for solving the classical 2×2 matrix games without saddle points, the optimal strategies for the players I and II and the value of the 2×2 interval-valued matrix game $\bar{A}' = ([a'_{Lij}, a'_{Rij}])_{2\times 2}$ are obtained as follows [25, 26]:

$$\mathbf{y}'^* = (y_1'^*, y_2'^*)^{\mathrm{T}} = \left(\frac{a'_{L22} - a'_{L21}}{a'_{L11} + a'_{L22} - a'_{L12} - a'_{L21}}, \frac{a'_{L11} - a'_{L12}}{a'_{L11} + a'_{L22} - a'_{L12} - a'_{L21}}\right)^{\mathrm{T}},$$

$$\mathbf{z}'^* = (z_1'^*, z_2'^*)^{\mathrm{T}} = \left(\frac{a'_{L22} - a'_{L12}}{a'_{L11} + a'_{L22} - a'_{L12} - a'_{L21}}, \frac{a'_{L11} - a'_{L21}}{a'_{L11} + a'_{L22} - a'_{L12} - a'_{L21}}\right)^{\mathrm{T}}$$

and

$$\bar{v}'^* = [v_L'^*, v_R'^*] = \left[\frac{a'_{L11}a'_{L22} - a'_{L12}a'_{L21}}{a'_{L11} + a'_{L22} - a'_{L12} - a'_{L21}}, \frac{a'_{R11}a'_{R22} - a'_{R12}a'_{R21}}{a'_{R11} + a'_{R22} - a'_{R12} - a'_{R21}}\right],$$

respectively.

According to Theorem 1.3, the optimal strategies $\mathbf{y}^*$ and $\mathbf{z}^*$ for the players I and II in the 2×2 interval-valued matrix game $\bar{\mathbf{A}}'$ are the same as those for the players I and II in the 2×2 interval-valued matrix game $\bar{\mathbf{A}} = ([a_{Lij}, a_{Rij}])_{2\times 2}$, i.e.,

$$\mathbf{y}^* = \mathbf{y}'^* = \left(\frac{a'_{L22} - a'_{L21}}{a'_{L11} + a'_{L22} - a'_{L12} - a'_{L21}}, \frac{a'_{L11} - a'_{L12}}{a'_{L11} + a'_{L22} - a'_{L12} - a'_{L21}}\right)^{\mathrm{T}},$$

and

$$\mathbf{z}^* = \mathbf{z}'^* = \left(\frac{a'_{L22} - a'_{L12}}{a'_{L11} + a'_{L22} - a'_{L12} - a'_{L21}}, \frac{a'_{L11} - a'_{L21}}{a'_{L11} + a'_{L22} - a'_{L12} - a'_{L21}}\right)^{\mathrm{T}}.$$

Also the value $\bar{v}^*$ of the 2×2 interval-valued matrix game $\bar{\mathbf{A}}$ is obtained as follows

$$\bar{v}^* = \mu\bar{v}'^* = \left[\mu\frac{a'_{L11}a'_{L22} - a'_{L12}a'_{L21}}{a'_{L11} + a'_{L22} - a'_{L12} - a'_{L21}}, \mu\frac{a'_{R11}a'_{R22} - a'_{R12}a'_{R21}}{a'_{R11} + a'_{R22} - a'_{R12} - a'_{R21}}\right].$$

Using $a'_{Rij} = a_{Rij}/\mu$ and $a'_{Lij} = a_{Lij}/\mu$ $(i,j = 1,2)$, the expression of the above optimal strategies $\mathbf{y}^*$ and $\mathbf{z}^*$ for the players I and II and the value $\bar{v}^*$ of the 2×2 interval-valued matrix game $\bar{\mathbf{A}}$ can be rewritten as Eqs. (1.15)–(1.17). Thus, we have finished the proof of Corollary 1.1.

Remark 1.1 If $a_{Rij} = a_{Lij} + \mu$ $(i,j = 1,2)$, then Eqs. (1.15)–(1.17) can be directly obtained through solving the systems of equations as follows:

$$\begin{cases} a_{L11}y_1 + a_{L21}(1 - y_1) = a_{L12}y_1 + a_{L22}(1 - y_1) \\ a_{R11}y_1 + a_{R21}(1 - y_1) = a_{R12}y_1 + a_{R22}(1 - y_1) \end{cases}$$

and

$$\begin{cases} a_{L11}z_1 + a_{L12}(1 - z_1) = a_{L21}z_1 + a_{L22}(1 - z_1) \\ a_{R11}z_1 + a_{R12}(1 - z_1) = a_{R21}z_1 + a_{R22}(1 - z_1), \end{cases}$$

respectively.

It is easy to see from Corollary 1.1 that Eqs. (1.15) and (1.16) could provide optimal strategies for the players I and II in the 2 × 2 interval-valued matrix game $\bar{A} = ([a_{Lij}, a_{Rij}])_{2\times 2}$ only if its intervals $[a_{Lij}, a_{Rij}]$ satisfy the conditions: $a_{Rij} = a_{Lij} + \mu$ $(i,j = 1,2)$, i.e., all the intervals are of the identical length μ. In other words, the solution provided by Eqs. (1.15)–(1.17) is a solution of the 2 × 2 interval-valued matrix game $\bar{A}'$ whereas it is not always a solution of the 2 × 2 interval-valued matrix game $\bar{A}$.

Example 1.3 Let us consider the 2 × 2 interval-valued matrix game $\bar{A}_2$, whose interval-valued payoff matrix is given as follows:

$$\bar{A}_2 = ([a_{Lij}, a_{Rij}])_{2\times 2} = \begin{matrix} \\ \delta_1 \\ \delta_2 \end{matrix} \begin{matrix} \beta_1 \quad\quad \beta_2 \\ \begin{pmatrix} [2,3] & [6,16] \\ [3,6] & [4,14] \end{pmatrix} \end{matrix}.$$

It is easy to verify that

$$\vee_{\psi_{1\le i\le 2}} \{ \wedge_{\psi_{1\le j\le 2}} \{[a_{Lij}, a_{Rij}]\}\} = \wedge_{\psi_{1\le j\le 2}} \{ \vee_{\psi_{1\le i\le 2}} \{[a_{Lij}, a_{Rij}]\}\} = [a_{L21}, a_{R21}] = [3,6].$$

Therefore, the interval-valued matrix game $\bar{A}_2$ has a saddle point (2,1) or (δ_2, β_1), i.e., the optimal strategies for the players I and II are the pure strategies δ_2 and β_1, respectively. The value of the interval-valued matrix game $\bar{A}_2$ is $\bar{v}^* = [3,6]$.

Obviously, the length of the intervals in the interval-valued payoff matrix $\bar{A}_2$ given in Example 1.3 is not identical. As a result, Corollary 1.1 is inapplicable. However, if we still employ a similar transform method, i.e.,

$$[a'_{Lij}, a'_{Rij}] = [\frac{a_{Lij}}{\mu_{ij}}, \frac{a_{Rij}}{\mu_{ij}}], \tag{1.18}$$

where $a_{Rij} = a_{Lij} + \mu_{ij}$ and $\mu_{ij} > 0$ $(i,j = 1,2)$, then the interval-valued payoff matrix $\bar{A}_2$ is converted into:

$$\bar{A}'_2 = ([a'_{Lij}, a'_{Rij}])_{2\times 2} = \begin{matrix} \\ \delta_1 \\ \delta_2 \end{matrix} \begin{matrix} \beta_1 \quad\quad \beta_2 \\ \begin{pmatrix} [2,3] & [0.6, 1.6] \\ [1,2] & [0.4, 1.4] \end{pmatrix} \end{matrix}.$$

It is easily seen that the interval-valued matrix game $\bar{A}'_2$ has a saddle point (1, 2) or (δ_1, β_2) due to

$$\vee_{\psi_{1\le i\le 2}} \{ \wedge_{\psi_{1\le j\le 2}} \{[a'_{Lij}, a'_{Rij}]\}\} = \wedge_{\psi_{1\le j\le 2}} \{ \vee_{\psi_{1\le i\le 2}} \{[a'_{Lij}, a'_{Rij}]\}\} = [a'_{L12}, a'_{R12}] = [0.6, 1.6].$$

Namely, the optimal strategies for the players I and II are the pure strategies δ_1 and β_2, respectively, which are remarkably different from those for the players I and II in the interval-valued matrix game $\bar{A}_2$. The value of the interval-valued matrix game $\bar{A}_2'$ is $\bar{v}'^* = [0.6, 1.6]$ which corresponds to the interval $[a_{L12}, a_{R12}] = [6, 16]$ in the original interval-valued matrix game $\bar{A}_2$. As a result, the interval-valued matrix game $\bar{A}_2$ has two different values. This shows that $[a_{L12}, a_{R12}] = [6, 16]$ obtained through using Eq. (1.18) is not the value of the interval-valued matrix game $\bar{A}_2$. In other words, the transform method [i.e., Eq. (1.18)] is not always effective for solving any 2 × 2 interval-valued matrix game.

Example 1.4 Let us consider the 2 × 2 interval-valued matrix game $\bar{A}_3$ with the interval-valued payoff matrix as follows:

$$\bar{A}_3 = ([a_{Lij}, a_{Rij}])_{2\times 2} = \begin{matrix} \\ \delta_1 \\ \delta_2 \end{matrix} \begin{matrix} \begin{matrix} \beta_1 & \beta_2 \end{matrix} \\ \begin{pmatrix} [2, 3] & [6, 16] \\ [4, 14] & [3, 9] \end{pmatrix} \end{matrix}.$$

It is easily verified that

$$\underset{1\le i\le 2}{\vee_\psi}\left\{\underset{1\le j\le 2}{\wedge_\psi}\{[a_{Lij}, a_{Rij}]\}\right\} = [a_{L22}, a_{R22}] = [3, 9] \neq \underset{1\le j\le 2}{\wedge_\psi}\left\{\underset{1\le i\le 2}{\vee_\psi}\{[a_{Lij}, a_{Rij}]\}\right\}$$
$$= [a_{L21}, a_{R21}] = [4, 14].$$

Therefore, the interval-valued matrix game $\bar{A}_3$ has no saddle points in the sense of the pure strategies.

Using Eq. (1.18), the interval-valued payoff matrix $\bar{A}_3$ is converted into:

$$\bar{A}_3' = ([a'_{Lij}, a'_{Rij}])_{2\times 2} = \begin{matrix} \\ \delta_1 \\ \delta_2 \end{matrix} \begin{matrix} \begin{matrix} \beta_1 & \beta_2 \end{matrix} \\ \begin{pmatrix} [2, 3] & [0.6, 1.6] \\ [0.4, 1.4] & [0.5, 1.5] \end{pmatrix} \end{matrix}.$$

It is easy to see that the interval-valued matrix game $\bar{A}_3'$ has a saddle point (1, 2) or (δ_1, β_2) due to

$$\underset{1\le i\le 2}{\vee_\psi}\left\{\underset{1\le j\le 2}{\wedge_\psi}\{[a'_{Lij}, a'_{Rij}]\}\right\} = \underset{1\le j\le 2}{\wedge_\psi}\left\{\underset{1\le i\le 2}{\vee_\psi}\{[a'_{Lij}, a'_{Rij}]\}\right\} = [a'_{L12}, a'_{R12}] = [0.6, 1.6].$$

Thus, the optimal strategies for the players I and II are the pure strategies δ_1 and β_2, respectively.

It is concluded that Corollary 1.1 is not effective if one of the values μ_{ij} is not equal to the common constant, i.e., the intervals are not of the identical length, where $\mu_{ij} = a_{Rij} - a_{Lij}$ $(i, j = 1, 2)$. In this case, Eq. (1.18) cannot ensure that 2 × 2 interval-valued matrix games $\bar{A} = ([a_{Lij}, a_{Rij}])_{2\times 2}$ and $\bar{A}' = ([a'_{Lij}, a'_{Rij}])_{2\times 2}$ have identical optimal strategies for the players I and II.

1.4 Acceptability-Degree-Based Linear Programming Models of Interval-Valued Matrix Games

In this section, let us still consider the interval-valued matrix game $\bar{\boldsymbol{A}}$ stated as in Sect. 1.3.2. Firstly, we introduce the concepts of acceptability degrees of interval comparison. Then, we propose the models and method of any interval-valued matrix game.

1.4.1 Concepts of Acceptability Degrees of Interval Comparison and Properties

Interval comparison or ranking plays an important role in solving interval-valued matrix games. It is a difficult problem [22, 24, 29]. In fact, in terms of the fuzzy set introduced by Zadeh [30], the statement "an interval $\bar{a}$ is not greater than an interval $\bar{b}$", which is denoted by $\bar{a} \leq_I \bar{b}$, may be regarded as a fuzzy relation between $\bar{a}$ and $\bar{b}$. Collins and Hu [24, 31] defined a fuzzy partial order relation for intervals through taking full account of the inclusion and/or overlap relations between intervals, depicted as in Fig. 1.1.

Definition 1.3 Let $\bar{a} = [a_L, a_R]$ and $\bar{b} = [b_L, b_R]$ be arbitrary intervals. The premise "$\bar{a} \leq_I \bar{b}$" is regarded as a fuzzy set, whose membership function is defined as follows:

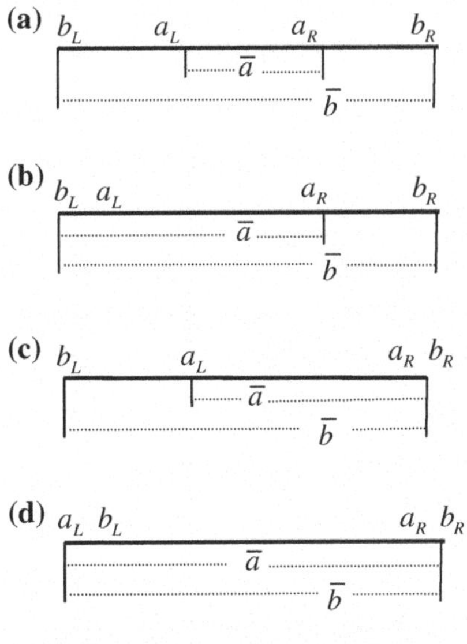

Fig. 1.1 Inclusion and/or overlap relations between two intervals

$$\varphi(\bar{a} \leq_I \bar{b}) = \begin{cases} 1 & \text{if } a_R < b_L \\ 1^- & \text{if } a_L < b_L \leq a_R < b_R \\ \dfrac{b_R - a_R}{2(w(\bar{b}) - w(\bar{a}))} & \text{if } b_L \leq a_L \leq a_R \leq b_R \text{ and } w(\bar{b}) > w(\bar{a}) \\ 0.5 & \text{if } w(\bar{a}) = w(\bar{b}) \text{ and } a_L = b_L, \end{cases}$$

where "1^-" is a fuzzy number being less than 1, which indicates the fact that the interval $\bar{a}$ is weakly not greater than the interval $\bar{b}$.

Obviously, $0 \leq \varphi(\bar{a} \leq_I \bar{b}) \leq 1$. Thus, $\varphi(\bar{a} \leq_I \bar{b})$ may be interpreted as the acceptability degree of the premise (or order relation) $\bar{a} \leq_I \bar{b}$. If $\varphi(\bar{a} \leq_I \bar{b}) = 0$, then the premise $\bar{a} \leq_I \bar{b}$ is not accepted. If $0 < \varphi(\bar{a} \leq_I \bar{b}) < 1$, then the player accepts the premise $\bar{a} \leq_I \bar{b}$ with different satisfactory degrees between 0 and 1. If $\varphi(\bar{a} \leq_I \bar{b}) = 1$, then the player is absolutely satisfied with the premise $\bar{a} \leq_I \bar{b}$. Namely, the player completely believes that the premise $\bar{a} \leq_I \bar{b}$ is true.

The symbol "$\leq_I$" is an interval version of the order relation "≤" in the real number set R and has the linguistic interpretation "essentially not greater than".

$\bar{a} <_I \bar{b}$ if and only if $\bar{a} \leq_I \bar{b}$ and $\bar{a} \neq \bar{b}$. The symbol "$<_I$" is an interval version of the order relation "<" in the real number set R and has the linguistic interpretation "essentially smaller than".

Analogously, the statement "an interval $\bar{a}$ is not less than an interval $\bar{b}$", denoted by $\bar{a} \geq_I \bar{b}$, can be defined as follows.

Definition 1.4 Let $\bar{a} = [a_L, a_R]$ and $\bar{b} = [b_L, b_R]$ be arbitrary intervals. The premise "$\bar{a} \geq_I \bar{b}$" is regarded as a fuzzy set, whose membership function is defined as $\varphi(\bar{a} \geq_I \bar{b}) = 1 - \varphi(\bar{a} \leq_I \bar{b})$, i.e.,

$$\varphi(\bar{a} \geq_I \bar{b}) = \begin{cases} 0 & \text{if } a_R < b_L \\ 0^- & \text{if } a_L < b_L \leq a_R < b_R \\ \dfrac{a_L - b_L}{2(w(\bar{b}) - w(\bar{a}))} & \text{if } b_L \leq a_L \leq a_R \leq b_R \text{ and } w(\bar{b}) > w(\bar{a}) \\ 0.5 & \text{if } w(\bar{a}) = w(\bar{b}) \text{ and } a_L = b_L. \end{cases}$$

It is easy to prove that the fuzzy ranking index φ is continuous except a single special case, i.e., $a_L = b_L$ and $w(\bar{a}) = w(\bar{b})$. Moreover, it is easily derived from Definitions 1.3 and 1.4 that there are some useful properties [27], which can be summarized as in Theorem 1.4 as follows.

Theorem 1.4 *For any intervals $\bar{a}$, $\bar{b}$, and $\bar{c}$, then*

(1) $0 \leq \varphi(\bar{a} \leq_I \bar{b}) \leq 1$ *or* $0 \leq \varphi(\bar{a} \geq_I \bar{b}) \leq 1$;
(2) $\varphi(\bar{a} \leq_I \bar{a}) = 0.5$ *or* $\varphi(\bar{a} \geq_I \bar{a}) = 0.5$;
(3) $\varphi(\bar{a} \leq_I \bar{b}) + \varphi(\bar{a} \geq_I \bar{b}) = 1$;
(4) *if* $\varphi(\bar{a} \leq_I \bar{b}) \geq 0.5$ *and* $\varphi(\bar{b} \leq_I \bar{c}) \geq 0.5$, *then* $\varphi(\bar{a} \leq_I \bar{c}) \geq 0.5$; *or if* $\varphi(\bar{a} \leq_I \bar{b}) \leq 0.5$ *and* $\varphi(\bar{b} \leq_I \bar{c}) \leq 0.5$, *then* $\varphi(\bar{a} \leq_I \bar{c}) \leq 0.5$.

Proof According to Definitions 1.3 and 1.4, we can easily prove that the conclusions are valid (omitted).

Similarly, $\bar{a} >_I \bar{b}$ if and only if $\bar{a} \geq_I \bar{b}$ and $\bar{a} \neq \bar{b}$. Also $\bar{a} =_I \bar{b}$ if and only if $\bar{a} \leq_I \bar{b}$ and $\bar{a} \geq_I \bar{b}$. The symbols "$\geq_I$", "$>_I$", and "$=_I$" are the interval versions of the order relations "≥", ">", and "=" in the real number set R and have the linguistic interpretation "essentially not smaller than", "essentially greater than", and "essentially being equal to", respectively.

Thus, "$\geq_I$" and "$\leq_I$" establish fuzzy partial orders for intervals. Definitions 1.3 and 1.4 may provide quantitative methods to determine the exact degree of membership (or satisfactory degree) for ranking (or comparing) two intervals. In this section, the fuzzy ranking index φ is used to define satisfactory crisp equivalent forms of interval valued inequality relations.

1.4.2 Interval-Valued Mathematical Programming Models and Satisfactory Equivalent Forms

We introduce the following definitions and equivalent forms, which are used in the sequent Sect. 1.4.3.

Definition 1.5 A satisfactory crisp equivalent form of an interval-valued inequality $\bar{a}x \leq_I \bar{b}$ is defined as follows:

$$\begin{cases} a_R x \leq b_R \\ \varphi(\bar{a}x \geq_I \bar{b}) \leq \varepsilon, \end{cases}$$

where $x \in \mathrm{R}$ is a variable and $\varepsilon \in [0, 1]$ represents the acceptance degree of the interval-valued inequality constraint which may be allowed to violate. Similarly, a satisfactory crisp equivalent form of an interval-valued inequality $\bar{a}x \geq_I \bar{b}$ is defined as follows:

$$\begin{cases} a_L x \geq b_L \\ \varphi(\bar{a}x \leq_I \bar{b}) \leq \varepsilon. \end{cases}$$

Ishibuchi and Tanaka [32] gave definitions of the maximization and minimization problems with the interval-valued objective functions, which are reviewed in Definitions 1.6 and 1.7 as follows.

Definition 1.6 Let $\bar{a} = [a_L, a_R]$ be an arbitrary interval. The maximization problem with the interval-valued objective function is described as follows:

$$\max\{\bar{a} | \bar{a} \in \Omega_1\},$$

which is equivalent to the bi-objective mathematical programming model as follows:

$$\begin{aligned} &\max\{a_L\} \\ &\max\{m(\bar{a})\} \\ &\text{s.t.} \begin{cases} \bar{a} \in \Omega_1 \\ a_L \text{ and } a_R \text{ unrestricted in sign,} \end{cases} \end{aligned}$$

where Ω_1 is the set of constraints in which the variable $\bar{a}$ should be satisfied according to requirements in the real situations.

Definition 1.7 Let $\bar{a} = [a_L, a_R]$ be an arbitrary interval. The minimization problem with the interval-valued objective function is described as follows:

$$\min\{\bar{a}|\bar{a} \in \Omega_2\},$$

which is equivalent to the bi-objective mathematical programming model as follows:

$$\begin{aligned} &\max\{a_R\} \\ &\max\{m(\bar{a})\} \\ &\text{s.t.} \begin{cases} \bar{a} \in \Omega_2 \\ a_L \text{ and } a_R \text{ unrestricted in sign,} \end{cases} \end{aligned}$$

where Ω_2 is the set of constraints in which the variable $\bar{a}$ should be satisfied according to requirements in the real situations.

In the next subsection, Definitions 1.3–1.7 will be used to transform corresponding interval-interval programming models of any interval-valued matrix game into bi-objective linear programming models.

1.4.3 Auxiliary Linear Programming Models of Interval-Valued Matrix Games

For the interval-valued matrix game $\bar{\boldsymbol{A}}$ stated as in Sect. 1.3.2, due to the player I being a maximizing player, therefore the player II is interested in finding a mixed strategy $\boldsymbol{z} \in Z$ so as to minimize $\bar{E}(\boldsymbol{y}, \boldsymbol{z})$ in the sense of the fuzzy ranking index φ given by Definitions 1.3 and 1.4, denoted by $\min_{\boldsymbol{z}\in Z}\{\bar{E}(\boldsymbol{y}, \boldsymbol{z})\}$ for short, where $\bar{E}(\boldsymbol{y}, \boldsymbol{z}) = \boldsymbol{y}^{\mathrm{T}}\bar{\boldsymbol{A}}\boldsymbol{z}$ is the interval-valued expected payoff of the player I. Hence, the player I should choose a mixed strategy $\boldsymbol{y} \in Y$ that maximizes $\min_{\boldsymbol{z}\in Z}\{\bar{E}(\boldsymbol{y}, \boldsymbol{z})\}$ of the player II (in the sense of the fuzzy ranking index φ), i.e.,

$$\bar{\upsilon}'^{*} = \max_{\boldsymbol{y}\in Y} \min_{\boldsymbol{z}\in Z}\{\bar{E}(\boldsymbol{y}, \boldsymbol{z})\},$$

which is called the player I's gain-floor.

Similarly, due to the player II being a minimizing player, the player I is interested in finding a mixed strategy $\boldsymbol{y} \in Y$ so as to maximize $\bar{E}(\boldsymbol{y}, \boldsymbol{z})$ in the sense of the fuzzy ranking index φ given by Definitions 1.3 and 1.4, denoted by $\max_{\boldsymbol{y} \in Y}\{\bar{E}(\boldsymbol{y}, \boldsymbol{z})\}$ for short. Thus, the player II should choose a mixed strategy $\boldsymbol{z} \in Z$ that minimizes $\max_{\boldsymbol{y} \in Y}\{\bar{E}(\boldsymbol{y}, \boldsymbol{z})\}$ of the player I, i.e.,

$$\bar{\omega}'^{*} = \min_{\boldsymbol{z} \in Z} \max_{\boldsymbol{y} \in Y}\{\bar{E}(\boldsymbol{y}, \boldsymbol{z})\},$$

which is called the player II's loss-ceiling.

Obviously, the player I's gain-floor and the player II's loss-ceiling should be intervals. Therefore, in a similar way to Definitions 1.1 and 1.2, we can define the solution of any interval-valued matrix game in the above fuzzy partial orders over intervals.

Definition 1.8 Let $\bar{\upsilon} = [\upsilon_L, \upsilon_R]$ and $\bar{\omega} = [\omega_L, \omega_R]$ be two intervals. Assume that there exist mixed strategies $\boldsymbol{y}^* \in Y$ and $\boldsymbol{z}^* \in Z$. If $(\boldsymbol{y}^*, \boldsymbol{z}^*, \bar{\upsilon}, \bar{\omega})$ satisfies both $\boldsymbol{y}^{*\mathrm{T}} \bar{\boldsymbol{A}} \boldsymbol{z} \geq_I \bar{\upsilon}$ and $\boldsymbol{y}^{\mathrm{T}} \bar{\boldsymbol{A}} \boldsymbol{z}^* \leq_I \bar{\omega}$ for any mixed strategies $\boldsymbol{y} \in Y$ and $\boldsymbol{z} \in Z$, then $(\boldsymbol{y}^*, \boldsymbol{z}^*, \bar{\upsilon}, \bar{\omega})$ is called a reasonable solution of the interval-valued matrix game $\bar{\boldsymbol{A}}$; $\bar{\upsilon}$ and $\bar{\omega}$ are called reasonable values for the players I and II, $\boldsymbol{y}^*$ and $\boldsymbol{z}^*$ are called reasonable strategies for the players I and II, respectively.

The sets of reasonable values $\bar{\upsilon}$ and $\bar{\omega}$ for the players I and II are denoted by U and W, respectively.

Definition 1.9 Assume that there exist two reasonable values $\bar{\upsilon}^* \in U$ and $\bar{\omega}^* \in W$. If there do not exist reasonable values $\bar{\upsilon}' \in U$ ($\bar{\upsilon}' \neq \bar{\upsilon}^*$) and $\bar{\omega}' \in W$ ($\bar{\omega}' \neq \bar{\omega}^*$) such that they satisfy both $\bar{\upsilon}' \geq_I \bar{\upsilon}^*$ and $\bar{\omega}' \leq_I \bar{\omega}^*$, then $(\boldsymbol{y}^*, \boldsymbol{z}^*, \bar{\upsilon}^*, \bar{\omega}^*)$ is called a solution of the interval-valued matrix game $\bar{\boldsymbol{A}}$; $\boldsymbol{y}^*$ is called an optimal (or a maximin) strategy for the player I and $\boldsymbol{z}^*$ is called an optimal (or a minimax) strategy for the player II; $\bar{\upsilon}^*$ and $\bar{\omega}^*$ are called the player I's gain-floor and the player II's loss-ceiling, respectively.

We can easily draw a conclusion which is similar to Eq. (1.12).

Theorem 1.5 *For any interval-valued matrix game* $\bar{\boldsymbol{A}}$, *then* $\bar{\upsilon}'^{*} \leq_I \bar{\omega}'^{*}$.

Proof For any mixed strategy $\boldsymbol{y} \in Y$, it directly follows that

$$\min_{\boldsymbol{z} \in Z}\{\bar{E}(\boldsymbol{y}, \boldsymbol{z})\} \leq_I \bar{E}(\boldsymbol{y}, \boldsymbol{z}).$$

Similarly, for any mixed strategy $\boldsymbol{z} \in Z$, we have

$$\max_{\boldsymbol{y} \in Y}\{\bar{E}(\boldsymbol{y}, \boldsymbol{z})\} \geq_I \bar{E}(\boldsymbol{y}, \boldsymbol{z}).$$

Thus, for any mixed strategies $\boldsymbol{y} \in Y$ and $\boldsymbol{z} \in Z$, we obtain

$$\min_{z \in Z}\{\bar{E}(\boldsymbol{y}, \boldsymbol{z})\} \leq_I \max_{y \in Y}\{\bar{E}(\boldsymbol{y}, \boldsymbol{z})\},$$

which infers that

$$\min_{z \in Z}\{\bar{E}(\boldsymbol{y}, \boldsymbol{z})\} \leq_I \min_{z \in Z} \max_{y \in Y}\{\bar{E}(\boldsymbol{y}, \boldsymbol{z})\}.$$

Hence, we have

$$\max_{y \in Y} \min_{z \in Z}\{\bar{E}(\boldsymbol{y}, \boldsymbol{z})\} \leq_I \min_{z \in Z} \max_{y \in Y}\{\bar{E}(\boldsymbol{y}, \boldsymbol{z})\},$$

i.e., $\bar{\upsilon}'^* \leq_I \bar{\omega}'^*$. Thus, we have completed the proof of Theorem 1.5.

Theorem 1.5 means that the player I's gain-floor "essentially cannot exceed" the player II's loss-ceiling. Furthermore, it is easy to see from Definitions 1.8 and 1.9 and Theorem 1.5 that $\bar{\upsilon}^* = \bar{\upsilon}'^*$ and $\bar{\omega}^* = \bar{\omega}'^*$ if $(\boldsymbol{y}^*, \boldsymbol{z}^*, \bar{\upsilon}^*, \bar{\omega}^*)$ is a solution of the interval-valued matrix game $\bar{\boldsymbol{A}}$.

In this section, we focus on developing a methodology for solving any interval-valued matrix game $\bar{\boldsymbol{A}}$. Firstly, we introduce Lemma 1.1 as follows.

Lemma 1.1

(1) Assume that there exists a set $\{\bar{c}_1, \bar{c}_2, \ldots, \bar{c}_n\}$ of n intervals, where each $\bar{c}_j$ $(j = 1, 2, \ldots, n)$ is an interval. Then, the following equality is valid

$$\min_{z \in Z}\{\sum_{j=1}^{n} \bar{c}_j z_j\} = \min_{1 \leq j \leq n}\{\bar{c}_j\};$$

(2) Assume that there exists a set $\{\bar{d}_1, \bar{d}_2, \ldots, \bar{d}_m\}$ of m intervals, where each $\bar{d}_i$ $(i = 1, 2, \ldots, m)$ is an interval. Then, the following equality is valid

$$\max_{y \in Y}\{\sum_{i=1}^{m} \bar{d}_i y_i\} = \max_{1 \leq i \leq m}\{\bar{d}_i\}.$$

Proof

(1) Using the fuzzy ranking index φ given by Definitions 1.3 and 1.4, without loss of generality, we assume that

$$\min_{1 \leq j \leq n}\{\bar{c}_j\} = \bar{c}_l,$$

which directly infers that

$$\bar{c}_j \geq_I \bar{c}_l \; (j = 1, 2, \ldots, n).$$

For any mixed strategy $z \in Z$, due to $z_j \geq 0$ $(j = 1, 2, \ldots, n)$, we obtain

$$\bar{c}_j z_j \geq_I \bar{c}_1 z_j \quad (j = 1, 2, \ldots, n).$$

Summing the above n interval-valued inequalities, it follows that

$$\sum_{j=1}^{n} \bar{c}_j z_j \geq_I \sum_{j=1}^{n} \bar{c}_l z_j = \bar{c}_l$$

since $z_j \geq 0$ $(j = 1, 2, \ldots, n)$ and $\sum_{j=1}^{n} z_j = 1$. Thus, we have

$$\min_{z \in Z}\{\sum_{j=1}^{n} \bar{c}_j z_j\} \geq_I \bar{c}_l. \tag{1.19}$$

On the other hand, due to the fact that $z^0 = (0, 0, \ldots, 0, 1, 0, \ldots 0)^{\mathrm{T}}$ may be regarded as a special case of mixed strategies, where $z_l = 1$, and $z_j = 0$ $(j = 1, 2, \ldots, n; j \neq l)$, i.e., z^0 is the pure strategy β_l, we obtain

$$\bar{c}_l = \bar{c}_1 \times 0 + \bar{c}_2 \times 0 + \cdots + \bar{c}_l \times 1 + \cdots + \bar{c}_n \times 0 = \sum_{j=1}^{n} \bar{c}_j z_j^0 \geq_I \min_{z \in Z}\{\sum_{j=1}^{n} \bar{c}_j z_j\}.$$

Combining with Eq. (1.19), we have

$$\min_{z \in Z}\{\sum_{j=1}^{n} \bar{c}_j z_j\} = \bar{c}_l = \min_{1 \leq j \leq n}\{\bar{c}_j\}.$$

Thus, we have proven the case (1) of Lemma 1.1.

Analogously, we can easily prove the case (2) of Lemma 1.1 (omitted).

To simplify the computation procedure of interval-valued matrix games, we give Theorem 1.6 as follows.

Theorem 1.6 *For any interval-valued matrix game* $\bar{\mathbf{A}}$, *we have*

$$\max_{y \in Y} \min_{z \in Z}\{\bar{E}(\mathbf{y}, \mathbf{z})\} = \max_{y \in Y} \min_{1 \leq j \leq n}\{\sum_{i=1}^{m} \bar{a}_{ij} y_i\}$$

and

$$\min_{z \in Z} \max_{y \in Y}\{\bar{E}(\mathbf{y}, \mathbf{z})\} = \min_{z \in Z} \max_{1 \leq i \leq m}\{\sum_{j=1}^{n} \bar{a}_{ij} z_j\}.$$

Proof It easily follows from Lemma 1.1 that

$$\max_{y\in Y}\min_{z\in Z}\{\bar{E}(y,z)\} = \max_{y\in Y}\min_{z\in Z}\{\sum_{j=1}^{n}(\sum_{i=1}^{m}\bar{a}_{ij}y_i)z_j\} = \max_{y\in Y}\min_{1\le j\le n}\{\sum_{i=1}^{m}\bar{a}_{ij}y_i\}$$

and

$$\min_{z\in Z}\max_{y\in Y}\{\bar{E}(y,z)\} = \min_{z\in Z}\max_{y\in Y}\{\sum_{i=1}^{m}(\sum_{j=1}^{n}\bar{a}_{ij}z_j)y_i\} = \min_{z\in Z}\max_{1\le i\le m}\{\sum_{j=1}^{n}\bar{a}_{ij}z_j\}.$$

Hereby, we have completed the proof of Theorem 1.6.

According to Definitions 1.8 and 1.9 and Theorem 1.6, the solution $(y^*, z^*, \bar{\upsilon}^*, \bar{\omega}^*)$ of any interval-valued matrix game $\bar{A}$ can be generated by solving a pair of interval-valued programming models as follows:

$$\max\{[\upsilon_L, \upsilon_R]\}$$
$$\text{s.t.}\begin{cases} \sum_{i=1}^{m}[a_{Lij}, a_{Rij}]y_i \ge_I [\upsilon_L, \upsilon_R] \quad (j=1,2,\ldots,n) \\ \sum_{i=1}^{m} y_i = 1 \\ y_i \ge 0 \quad (i=1,2,\ldots,m) \\ \upsilon_L \text{ and } \upsilon_R \text{ unrestricted in sign} \end{cases} \tag{1.20}$$

and

$$\min\{[\omega_L, \omega_R]\}$$
$$\text{s.t.}\begin{cases} \sum_{j=1}^{n}[a_{Lij}, a_{Rij}]z_j \le_I [\omega_L, \omega_R] \quad (i=1,2,\ldots,m) \\ \sum_{j=1}^{n} z_j = 1 \\ z_j \ge 0 \quad (j=1,2,\ldots,n) \\ \omega_L \text{ and } \omega_R \text{ unrestricted in sign,} \end{cases} \tag{1.21}$$

respectively.

Equations (1.20) and (1.21) are generic interval-valued programming models which may be solved by different methods [33, 34]. In this section, interval-valued programming is made in the sense of Definitions 1.3–1.7. In the following, we focus on studying the solution method and procedure of Eqs. (1.20) and (1.21).

According to Definitions 1.3, 1.5, and 1.6, Eq. (1.20) can be transformed into the bi-objective programming model as follows:

$$\max\{\upsilon_L\}$$
$$\max\{\frac{\upsilon_L+\upsilon_R}{2}\}$$
$$\text{s.t.}\begin{cases}\sum_{i=1}^{m} a_{Lij}y_i \ge \upsilon_L \quad (j=1,2,\ldots,n)\\ \dfrac{\upsilon_R-\sum_{i=1}^{m} a_{Rij}y_i}{(\upsilon_R-\upsilon_L)-(\sum_{i=1}^{m} a_{Rij}y_i-\sum_{i=1}^{m} a_{Lij}y_i)} \le \varepsilon \quad (j=1,2,\ldots,n)\\ \upsilon_L \le \upsilon_R\\ \sum_{i=1}^{m} y_i = 1\\ y_i \ge 0 \quad (i=1,2,\ldots,m)\\ \upsilon_L \text{ and } \upsilon_R \text{ unrestricted in sign,}\end{cases}$$

which can be rewritten as the following bi-objective linear programming model:

$$\max\{\upsilon_L\}$$
$$\max\{\frac{\upsilon_L+\upsilon_R}{2}\}$$
$$\text{s.t.}\begin{cases}\sum_{i=1}^{m} a_{Lij}y_i \ge \upsilon_L \quad (j=1,2,\ldots,n)\\ (1-\varepsilon)\sum_{i=1}^{m} a_{Rij}y_i+\varepsilon\sum_{i=1}^{m} a_{Lij}y_i \ge (1-\varepsilon)\upsilon_R+\varepsilon\upsilon_L \quad (j=1,2,\ldots,n)\\ \upsilon_L \le \upsilon_R\\ \sum_{i=1}^{m} y_i = 1\\ y_i \ge 0 \quad (i=1,2,\ldots,m)\\ \upsilon_L \text{ and } \upsilon_R \text{ unrestricted in sign,}\end{cases} \tag{1.22}$$

where $\varepsilon \in [0,1]$ is given by the players *a priori*, which expresses the acceptance degree of the system of interval-valued inequalities which may be allowed to violate.

There are few standard ways of defining a solution for multi-objective programming. Normally, the concept of a Pareto optimal solution/efficient solution is commonly-used. Here, the weighted average method is used. As a result, Eq. (1.22) may be aggregated into the linear programming model as follows:

$$\max\left\{\frac{3\upsilon_L+\upsilon_R}{4}\right\}$$
$$\text{s.t.}\begin{cases}\sum_{i=1}^{m} a_{Lij}y_i \geq \upsilon_L \quad (j=1,2,\ldots,n)\\ (1-\varepsilon)\sum_{i=1}^{m} a_{Rij}y_i+\varepsilon\sum_{i=1}^{m} a_{Lij}y_i \geq (1-\varepsilon)\upsilon_R+\varepsilon\upsilon_L \ (j=1,2,\ldots,n)\\ \upsilon_L \leq \upsilon_R\\ \sum_{i=1}^{m} y_i=1\\ y_i \geq 0 \quad (i=1,2,\ldots,m)\\ \upsilon_L \text{ and } \upsilon_R \text{ unrestricted in sign,}\end{cases} \tag{1.23}$$

where y_i $(i=1,2,\cdots,m)$, υ_L, and υ_R are variables.

Using the existing simplex method for linear programming, an optimal solution of Eq. (1.23) can be obtained, denoted by $(\boldsymbol{y}^*, \upsilon_L^*, \upsilon_R^*)$.

It is not difficult to prove that $(\boldsymbol{y}^*, \bar{\upsilon}^*)$ is a Pareto optimal solution of Eq. (1.22), where $\bar{\upsilon}^*=[\upsilon_L^*, \upsilon_R^*]$ is an interval. Thus, the maximin (or optimal) mixed strategy $\boldsymbol{y}^*$ and the gain-floor $\bar{\upsilon}^*$ for the player I can be obtained.

In a similar consideration, according to Definitions 1.4, 1.5, and 1.7, Eq. (1.21) can be transformed into the bi-objective programming model as follows:

$$\min\{\omega_R\}$$
$$\min\left\{\frac{\omega_L+\omega_R}{2}\right\}$$
$$\text{s.t.}\begin{cases}\sum_{j=1}^{n} a_{Rij}z_j \leq \omega_R \quad (i=1,2,\ldots,m)\\ \dfrac{\sum_{j=1}^{n} a_{Lij}z_j-\omega_L}{(\omega_R-\omega_L)-(\sum_{j=1}^{n} a_{Rij}z_j-\sum_{j=1}^{n} a_{Lij}z_j)} \leq \varepsilon \quad (i=1,2,\ldots,m)\\ \omega_L \leq \omega_R\\ \sum_{j=1}^{n} z_j=1\\ z_j \geq 0 \quad (j=1,2,\ldots,n)\\ \omega_L \text{ and } \omega_R \text{ unrestricted in sign,}\end{cases}$$

which may be rewritten as the following bi-objective linear programming model:

$$
\begin{aligned}
&\min\{\omega_R\}\\
&\min\{\frac{\omega_L+\omega_R}{2}\}\\
&\text{s.t.}\begin{cases}
\sum\limits_{j=1}^{n} a_{Rij}z_j \le \omega_R \quad (i=1,2,\ldots,m)\\
(1-\varepsilon)\sum\limits_{j=1}^{n} a_{Lij}z_j+\varepsilon\sum\limits_{j=1}^{n} a_{Rij}z_j \le (1-\varepsilon)\omega_L+\varepsilon\omega_R \quad (i=1,2,\ldots,m)\\
\omega_L\le\omega_R\\
\sum\limits_{j=1}^{n} z_j=1\\
z_j\ge 0 \quad (j=1,2,\ldots,n)\\
\omega_L \text{ and } \omega_R \text{ unrestricted in sign,}
\end{cases}
\end{aligned}
\tag{1.24}
$$

where z_j $(j=1,2,\cdots,n)$, ω_L, and ω_R are variables, and $\varepsilon\in[0,1]$ is given *a priori*.

Analogously, using the weighted average method, Eq. (1.24) may be aggregated into the linear programming model as follows:

$$
\begin{aligned}
&\min\{\frac{3\omega_R+\omega_L}{4}\}\\
&\text{s.t.}\begin{cases}
\sum\limits_{j=1}^{n} a_{Rij}z_j \le \omega_R \quad (i=1,2,\cdots,m)\\
(1-\varepsilon)\sum\limits_{j=1}^{n} a_{Lij}z_j+\varepsilon\sum\limits_{j=1}^{n} a_{Rij}z_j \le (1-\varepsilon)\omega_L+\varepsilon\omega_R \quad (i=1,2,\cdots,m)\\
\omega_L\le\omega_R\\
\sum\limits_{j=1}^{n} z_j=1\\
z_j\ge 0 \quad (j=1,2,\ldots,n)\\
\omega_L \text{ and } \omega_R \text{ unrestricted in sign.}
\end{cases}
\end{aligned}
\tag{1.25}
$$

Applying the existing simplex method for linear programming, we can obtain an optimal solution of Eq. (1.25), denoted by $(z^*,\omega_L^*,\omega_R^*)$.

It is not difficult to prove that $(z^*,\bar{\omega}^*)$ is a Pareto optimal solution of Eq. (1.24), where $\bar{\omega}^*=[\omega_L^*,\omega_R^*]$ is an interval. Therefore, the minimax (or optimal) strategy z^* and the loss-ceiling $\bar{\omega}^*$ for the player II can be obtained.

Obviously, if all intervals $\bar{a}=[a_{Lij},a_{Rij}]$ in the interval-valued payoff matrix $\bar{\boldsymbol{A}}$ are real numbers, i.e., $a_{Lij}=a_{Rij}=a_{ij}$ $(i=1,2,\ldots,m;j=1,2,\ldots,n)$, then $\bar{\upsilon}$ and $\bar{\omega}$ are also real numbers, i.e., $\upsilon_L=\upsilon_R=\upsilon$ and $\omega_L=\omega_R=\omega$. Thus, Eqs. (1.23) and (1.25) are reduced to the linear programming models as follows:

$$\max\{\upsilon\}$$
$$\text{s.t.}\begin{cases}\sum\limits_{i=1}^{m} a_{ij}y_i \geq \upsilon \quad (j=1,2,\ldots,n)\\ \sum\limits_{i=1}^{m} y_i = 1\\ y_i \geq 0 \quad (i=1,2,\ldots,m)\\ \upsilon \text{ unrestricted in sign}\end{cases}$$

and

$$\min\{\omega\}$$
$$\text{s.t.}\begin{cases}\sum\limits_{j=1}^{n} a_{ij}z_j \leq \omega \quad (i=1,2,\ldots,m)\\ \sum\limits_{j=1}^{n} z_j = 1\\ z_j \geq 0 \quad (j=1,2,\ldots,n)\\ \omega \text{ unrestricted in sign,}\end{cases}$$

respectively, which are just about the linear programming models of the classical matrix game $\boldsymbol{A} = (a_{ij})_{m\times n}$.

1.4.4 Real Example Analysis of Market Share Problems

Suppose that there are two companies p_1 and p_2 aiming to enhance the market share of a product in a targeted market under the circumstance that the demand amount of the product in the targeted market is fixed basically. In other words, the market share of one company increases while the market share of another company decreases. The companies are considering about two options (or pure strategies) to increase the market share: advertisement (β_1) and reducing the price (β_2). The above problem may be regarded as a matrix game problem. Namely, the companies p_1 and p_2 are regarded as the players I and II, respectively. They may use the pure strategies β_1 and β_2. Due to a lack of information or imprecision of the available information, the managers of the companies usually are not able to exactly forecast the sales amount of the companies. Assume that the interval-valued payoff matrix for the company p_1 is given as follows:

$$\bar{\boldsymbol{A}}_2 = \begin{matrix} \\ \beta_1 \\ \beta_2 \end{matrix} \begin{matrix} \begin{matrix} \beta_1 & \qquad\qquad \beta_2 \end{matrix} \\ \begin{pmatrix} [175, 190] & [120, 158] \\ [80, 100] & [180, 190] \end{pmatrix} \end{matrix},$$

where [175, 190] in the interval-valued payoff matrix $\bar{\boldsymbol{A}}_2$ is an interval which indicates that the sales amount of the company p_1 varies within a range from 175 to 190 when the companies p_1 and p_2 use the pure strategy β_1 (advertisement) simultaneously. Other intervals in the interval-valued payoff matrix $\bar{\boldsymbol{A}}_2$ are explained similarly.

According to Eq. (1.23), the linear programming model is obtained as follows:

$$\max\{\frac{3\upsilon_L+\upsilon_R}{4}\}$$
$$\text{s.t.}\begin{cases} 175y_1+80y_2\geq\upsilon_L \\ 120y_1+180y_2\geq\upsilon_L \\ (1-\varepsilon)(190y_1+100y_2)+\varepsilon(175y_1+80y_2)\geq(1-\varepsilon)\upsilon_R+\varepsilon\upsilon_L \\ (1-\varepsilon)(158y_1+190y_2)+\varepsilon(120y_1+180y_2)\geq(1-\varepsilon)\upsilon_R+\varepsilon\upsilon_L \\ \upsilon_L\leq\upsilon_R \\ y_1+y_2=1 \\ y_1\geq 0,\ y_2\geq 0 \\ \upsilon_L \text{ and } \upsilon_R \text{ unrestricted in sign,} \end{cases} \tag{1.26}$$

where y_i $(i=1,2)$, υ_L, and υ_R are variables.

For some specific given values of the parameter $\varepsilon\in[0,1]$, solving Eq. (1.26) by using the simplex method for linear programming, the maximin (or optimal) strategies $\boldsymbol{y}^*$ and the gain-floors $\bar{\upsilon}^*=[\upsilon_L^*,\upsilon_R^*]$ for the company p_1 can be obtained, depicted as in Table 1.1.

Table 1.1 Solutions of the interval-valued matrix game $\bar{\boldsymbol{A}}_2$

ε	$\boldsymbol{y}^{*\mathrm{T}}$	$\bar{\upsilon}^*=[\upsilon_L^*,\upsilon_R^*]$	$\boldsymbol{z}^{*\mathrm{T}}$	$\bar{\omega}^*=[\omega_L^*,\omega_R^*]$
0.0	(0.645, 0.355)	[141.3, 158.1]	(0.387, 0.613)	[141.3, 170.4]
0.1	(0.645, 0.355)	[141.3, 158.1]	(0.377, 0.623)	[140.7, 170.1]
0.2	(0.645, 0.355)	[141.3, 158.1]	(0.367, 0.633)	[140.2, 169.7]
0.3	(0.645, 0.355)	[141.3, 158.1]	(0.356, 0.644)	[139.6, 169.4]
0.4	(0.695, 0.305)	[138.3, 167.8]	(0.344, 0.656)	[138.9, 169.0]
0.5	(0.681, 0.319)	[139.1, 168.2]	(0.332, 0.668)	[138.3, 168.6]
0.55	(0.677, 0.323)	[139.1, 168.2]	(0.326, 0.674)	[137.9, 168.4]
0.56	(0.681, 0.319)	[139.2, 168.2]	(0.325, 0.675)	[137.9, 168.4]
0.58	(0.679, 0.321)	[139.3, 168.3]	(0.322, 0.678)	[137.7, 168.3]
0.6	(0.677, 0.323)	[139.4, 168.3]	(0.320, 0.680)	[137.6, 168.2]
0.7	(0.669, 0.331)	[139.9, 168.6]	(0.306, 0.694)	[136.8, 167.8]
0.8	(0.660, 0.340)	[0, 730.4]	(0.292, 0.708)	[0, 201.4]
0.9	(0.653, 0.347)	[0, 1436.7]	(0.278, 0.722)	[0, 181.9]

Analogously, according to Eq. (1.25), the linear programming model can be obtained as follows:

$$\min\{\frac{3\omega_R+\omega_L}{4}\}$$
$$\text{s.t.}\begin{cases} 190z_1+158z_2\leq\omega_R \\ 100z_1+190z_2\leq\omega_R \\ (1-\varepsilon)(175z_1+120z_2)+\varepsilon(190z_1+158z_2)\leq(1-\varepsilon)\omega_L+\alpha\omega_R \\ (1-\varepsilon)(80z_1+180z_2)+\varepsilon(100z_1+190z_2)\leq(1-\varepsilon)\omega_L+\alpha\omega_R \\ \omega_L\leq\omega_R \\ z_1+z_2=1 \\ z_1\geq 0,\ z_2\geq 0 \\ \omega_L \text{ and } \omega_R \text{ unrestricted in sign,} \end{cases} \quad (1.27)$$

where z_j $(j=1,2)$, ω_L, and ω_R are variables.

For the specific given values of the parameter $\varepsilon\in[0,1]$, solving Eq. (1.27) by using the simplex method for linear programming, the minimax (or optimal) strategies z^* and the loss-ceilings $\bar{\omega}^*=[\omega_L^*,\omega_R^*]$ for the company p_2 can be obtained, also depicted as in Table 1.1.

According to Definition 1.3 and the values $\bar{\omega}^*$ and $\bar{\upsilon}^*$ in Table 1.1, we obtain $\varphi(\bar{\omega}^*\leq_I\bar{\upsilon}^*)=1$ for any value $\varepsilon\in[0.58,1]$, which implies that the companies p_1 and p_2 are absolutely satisfied with $\bar{\omega}^*\leq_I\bar{\upsilon}^*$. Namely, the loss-ceiling $\bar{\omega}^*$ of the company p_2 is absolutely not greater than the gain-floor $\bar{\upsilon}^*$ of the company p_1, which contradicts with Theorem 1.5. Thus, the optimal threshold of the acceptance degree ε should be equal to 0.58. In other words, if $\varepsilon>0.58$, then the third and fourth inequality constraints in Eqs. (1.26) and (1.27) are violated. Hence, the obtained results are not the solutions of the interval-valued matrix game $\bar{\boldsymbol{A}}_2$ although they are the solutions to Eqs. (1.26) and (1.27).

1.5 The Lexicographic Method of Interval-Valued Matrix Games

Stated as the above, Eqs. (1.22) and (1.24) are the auxiliary bi-objective linear programming models of the interval-valued matrix game $\bar{\boldsymbol{A}}$ in Sect. 1.3.2, which may be solved through using the existing methods for multi-objective programming. In this section, we develop a lexicographic method for solving Eqs. (1.22) and (1.24), which is summarized as follows [26].

According to Eqs. (1.22) and (1.24), the linear programming models can be constructed as follows:

$$\max\{\upsilon_L\}$$
$$\text{s.t.}\begin{cases} \sum_{i=1}^{m} a_{Lij}y_i \ge \upsilon_L \quad (j = 1, 2, \ldots, n) \\ (1-\varepsilon)\sum_{i=1}^{m} a_{Rij}y_i + \varepsilon\sum_{i=1}^{m} a_{Lij}y_i \ge (1-\varepsilon)\upsilon_R + \varepsilon\upsilon_L \quad (j = 1, 2, \ldots, n) \\ \upsilon_L \le \upsilon_R \\ \sum_{i=1}^{m} y_i = 1 \\ y_i \ge 0 \quad (i = 1, 2, \ldots, m) \\ \upsilon_L \text{ and } \upsilon_R \text{ unrestricted in sign} \end{cases} \tag{1.28}$$

and

$$\min\{\omega_R\}$$
$$\text{s.t.}\begin{cases} \sum_{j=1}^{n} a_{Rij}z_j \le \omega_R \quad (i = 1, 2, \ldots, m) \\ (1-\varepsilon)\sum_{j=1}^{n} a_{Lij}z_j + \varepsilon\sum_{j=1}^{n} a_{Rij}z_j \le (1-\varepsilon)\omega_L + \varepsilon\omega_R \quad (i = 1, 2, \ldots, m) \\ \omega_L \le \omega_R \\ \sum_{j=1}^{n} z_j = 1 \\ z_j \ge 0 \quad (j = 1, 2, \ldots, n) \\ \omega_L \text{ and } \omega_R \text{ unrestricted in sign,} \end{cases} \tag{1.29}$$

respectively, where y_i $(i = 1, 2, \ldots, m)$, υ_L, υ_R, z_j $(j = 1, 2, \ldots, n)$, ω_L, and ω_R are variables, and the parameter $\varepsilon \in [0, 1]$ is given by the players *a priori*.

Solving Eqs. (1.28) and (1.29) by using the simplex method for linear programming, we obtain their optimal solutions, denoted by $(\boldsymbol{y}^0, \upsilon_L^0, \upsilon_R^0)$ and $(\boldsymbol{z}^0, \omega_L^0, \omega_R^0)$, respectively.

Hereby, again according to Eqs. (1.22) and (1.24), the linear programming models can be constructed as follows:

$$\max\{\frac{\upsilon_L + \upsilon_R}{2}\}$$

$$\text{s.t.}\begin{cases} \sum_{i=1}^{m} a_{Lij} y_i \geq \upsilon_L \quad (j = 1, 2, \ldots, n) \\ (1-\varepsilon)\sum_{i=1}^{m} a_{Rij} y_i + \varepsilon \sum_{i=1}^{m} a_{Lij} y_i \geq (1-\varepsilon)\upsilon_R + \varepsilon \upsilon_L \quad (j = 1, 2, \ldots, n) \\ \upsilon_L \leq \upsilon_R \\ \upsilon_L \geq \upsilon_L^0 \\ \sum_{i=1}^{m} y_i = 1 \\ y_i \geq 0 \quad (i = 1, 2, \ldots, m) \\ \upsilon_L \text{ and } \upsilon_R \text{ unrestricted in sign} \end{cases} \tag{1.30}$$

and

$$\min\{\frac{\omega_L + \omega_R}{2}\}$$

$$\text{s.t.}\begin{cases} \sum_{j=1}^{n} a_{Rij} z_j \leq \omega_R \quad (i = 1, 2, \ldots, m) \\ (1-\varepsilon)\sum_{j=1}^{n} a_{Lij} z_j + \varepsilon \sum_{j=1}^{n} a_{Rij} z_j \leq (1-\varepsilon)\omega_L + \varepsilon \omega_R \quad (i = 1, 2, \ldots, m) \\ \omega_L \leq \omega_R \\ \omega_R \leq \omega_R^0 \\ \sum_{j=1}^{n} z_j = 1 \\ z_j \geq 0 \quad (j = 1, 2, \ldots, n) \\ \omega_L \text{ and } \omega_R \text{ unrestricted in sign,} \end{cases} \tag{1.31}$$

respectively. Solving Eqs. (1.30) and (1.31) by using the simplex method for linear programming, we obtain their optimal solutions, denoted by $(\boldsymbol{y}^*, \upsilon_L^*, \upsilon_R^*)$ and $(\boldsymbol{z}^*, \omega_L^*, \omega_R^*)$, respectively.

It is not difficult to prove that $(\boldsymbol{y}^*, \upsilon_L^*, \upsilon_R^*)$ and $(\boldsymbol{z}^*, \omega_L^*, \omega_R^*)$ are the Pareto optimal solutions to Eqs. (1.22) and (1.24), respectively. Thus, $\boldsymbol{y}^*$ and $\bar{\upsilon}^* = [\upsilon_L^*, \upsilon_R^*]$ are the maximin (or optimal) strategy and the gain-floor for the player I, $\boldsymbol{z}^*$ and $\bar{\omega}^* = [\omega_L^*, \omega_R^*]$ are the minimax (or optimal) strategy and the loss-ceiling for the player II, respectively.

Example 1.5 Let us consider the interval-valued matrix game $\bar{A}_3$, whose interval-valued payoff matrix is adopted from Example 4 given by Li [26] as follows:

$$\bar{A}_3 = ([a_{Lij}, a_{Rij}])_{2\times 2} = \begin{matrix} \\ \delta_1 \\ \delta_2 \end{matrix} \begin{matrix} \beta_1 \qquad\qquad \beta_2 \\ \begin{pmatrix} [-3,-1] & [4,6] \\ [6,8] & [-7,-5] \end{pmatrix} \end{matrix}.$$

In the following, we solve this interval-valued matrix game $\bar{A}_3$ by using the above lexicographic method.

Taking $\varepsilon = 0$, which indicates that the inequality constraints are not allowed to violate, thus according to Eqs. (1.28) and (1.29), we obtain the linear programming models as follows:

$$\max\{\upsilon_L\}$$
$$\text{s.t.} \begin{cases} -3y_1 + 6y_2 \geq \upsilon_L \\ 4y_1 - 7y_2 \geq \upsilon_L \\ -y_1 + 8y_2 \geq \upsilon_R \\ 6y_1 - 5y_2 \geq \upsilon_R \\ \upsilon_L \leq \upsilon_R \\ y_1 + y_2 = 1 \\ y_1 \geq 0, y_2 \geq 0 \\ \upsilon_L \text{ and } \upsilon_R \text{ unrestricted in sign} \end{cases} \tag{1.32}$$

and

$$\min\{\omega_R\}$$
$$\text{s.t.} \begin{cases} -z_1 + 6z_2 \leq \omega_R \\ 8z_1 - 5z_2 \leq \omega_R \\ -3z_1 + 4z_2 \leq \omega_L \\ 6z_1 - 7z_2 \leq \omega_L \\ \omega_L \leq \omega_R \\ z_1 + z_2 = 1 \\ z_1 \geq 0, z_2 \geq 0 \\ \omega_L \text{ and } \omega_R \text{ unrestricted in sign,} \end{cases} \tag{1.33}$$

respectively.

Applying the simplex method for linear programming, we obtain the solutions of Eqs. (1.32) and (1.33) as follows:

$$(\mathbf{y}^0, \upsilon_L^0, \upsilon_R^0) = (y_1^0, y_2^0, \upsilon_L^0, \upsilon_R^0) = (\frac{13}{20}, \frac{7}{20}, \frac{163}{20}, \frac{243}{20})$$

and

$$(\mathbf{z}^0, \omega_L^0, \omega_R^0) = (z_1^0, z_2^0, \omega_L^0, \omega_R^0) = (\frac{11}{20}, \frac{9}{20}, \frac{163}{20}, \frac{243}{20}),$$

respectively.

According to Eqs. (1.30) and (1.31), and combining with $(\mathbf{y}^0, \upsilon_L^0, \upsilon_R^0)$ and $(\mathbf{z}^0, \omega_L^0, \omega_R^0)$, we obtain the linear programming models as follows:

$$\max\{\frac{\upsilon_L + \upsilon_R}{2}\}$$
$$\text{s.t.}\begin{cases} -3y_1 + 6y_2 \geq \upsilon_L \\ 4y_1 - 7y_2 \geq \upsilon_L \\ -y_1 + 8y_2 \geq \upsilon_R \\ 6y_1 - 5y_2 \geq \upsilon_R \\ \upsilon_L \leq \upsilon_R \\ \upsilon_L \geq 163/20 \\ y_1 + y_2 = 1 \\ y_1 \geq 0, y_2 \geq 0 \\ \upsilon_L \text{ and } \upsilon_R \text{ unrestricted in sign} \end{cases} \tag{1.34}$$

and

$$\min\{\frac{\omega_L + \omega_R}{2}\}$$
$$\text{s.t.}\begin{cases} -z_1 + 6z_2 \leq \omega_R \\ 8z_1 - 5z_2 \leq \omega_R \\ -3z_1 + 4z_2 \leq \omega_L \\ 6z_1 - 7z_2 \leq \omega_L \\ \omega_L \leq \omega_R \\ \omega_R \leq 243/20 \\ z_1 + z_2 = 1 \\ z_1 \geq 0, z_2 \geq 0 \\ \omega_L \text{ and } \omega_R \text{ unrestricted in sign,} \end{cases} \tag{1.35}$$

respectively.

Using the simplex method for linear programming, we obtain the solutions of Eqs. (1.34) and (1.35) as follows:

$$(\boldsymbol{y}^*, \upsilon_L^*, \upsilon_R^*) = (y_1^*, y_2^*, \upsilon_L^*, \upsilon_R^*) = (\frac{13}{20}, \frac{7}{20}, \frac{163}{20}, \frac{243}{20})$$

and

$$(\boldsymbol{z}^*, \omega_L^*, \omega_R^*) = (z_1^*, z_2^*, \omega_L^*, \omega_R^*) = (\frac{11}{20}, \frac{9}{20}, \frac{163}{20}, \frac{243}{20}),$$

respectively. Hence, the value of the interval-valued matrix game $\bar{\boldsymbol{A}}_3$ is obtained as follows:

$$\bar{v}^* = \sum_{i=1}^{2}\sum_{j=1}^{2} y_i^*[a_{Lij}, a_{Rij}]z_j^* = \left[\frac{3}{20}, \frac{43}{20}\right].$$

Thus, we obtain the optimal strategy $\boldsymbol{y}^* = (13/20, 7/20)^{\mathrm{T}}$ and the gain-floor $\bar{\upsilon}^* = [\upsilon_L^*, \upsilon_R^*] = [163/20, 243/20]$ for the player I, the optimal strategy $\boldsymbol{z}^* = (11/20, 9/20)^{\mathrm{T}}$ and the loss-ceiling $\bar{\omega}^* = [\omega_L^*, \omega_R^*] = [163/20, 243/20]$ for the player II as well as the value $\bar{v}^* = [3/20, 43/20]$ of the interval-valued matrix game $\bar{\boldsymbol{A}}_3$.

1.6 Primal-Dual Linear Programming Models of Interval-Valued Matrix Games

1.6.1 The Monotonicity of Values of Interval-Valued Matrix Games

Let us continue considering the interval-valued matrix game $\bar{\boldsymbol{A}}$ stated as in Sect. 1.3.2.

For any given values a_{ij} in the interval-valued payoffs $\bar{a}_{ij} = [a_{Lij}, a_{Rij}]$ $(i = 1, 2, \ldots, m; j = 1, 2, \ldots, n)$, a payoff matrix is denoted by $\boldsymbol{A} = (a_{ij})_{m \times n}$. It is easy to see from Eqs. (1.3) and (1.4) that the value v of the matrix game $\boldsymbol{A}$ for the player I is closely related to all values a_{ij}, i.e., entries in the payoff matrix $\boldsymbol{A}$. In other words, v is a function of the values a_{ij} $(i = 1, 2, \ldots, m; j = 1, 2, \ldots, n)$ in the interval-valued payoffs $\bar{a}_{ij}$, denoted by $v = \upsilon((a_{ij}))$ or $v = \upsilon(\boldsymbol{A})$. Similarly, the optimal (mixed) strategy $\boldsymbol{y}^* \in Y$ of the player I in the matrix game $\boldsymbol{A}$ is also a function of the values a_{ij} $(i = 1, 2, \ldots, m; j = 1, 2, \ldots, n)$, denoted by $\boldsymbol{y}^* = \boldsymbol{y}^*((a_{ij}))$ or $\boldsymbol{y}^* = \boldsymbol{y}^*(\boldsymbol{A})$.

In a similar way to the above analysis, according to Eqs. (1.6) and (1.7), the value μ and the optimal (mixed) strategy $\boldsymbol{z}^* \in Z$ for the player II in the matrix game

$\boldsymbol{A}$ are functions of the values a_{ij} $(i=1,2,\ldots,m;\ j=1,2,\ldots,n)$ in the interval-valued payoffs $\bar{a}_{ij}$, denoted by $\mu=\omega((a_{ij}))$ (or $\mu=\omega(\boldsymbol{A})$) and $z^*=z^*((a_{ij}))$ (or $z^*=z^*(\boldsymbol{A})$), respectively.

It is easy to see from Eqs. (1.3) and (1.4) that the value $v=\upsilon((a_{ij}))$ of the matrix game $\boldsymbol{A}$ for the player I is a non-decreasing function of the values a_{ij} $(i=1,2,\ldots,m;\ j=1,2,\ldots,n)$ in the interval-valued payoffs $\bar{a}_{ij}$. In fact, for any values a_{ij} and a'_{ij} in the interval-valued payoffs $\bar{a}_{ij}$ $(i=1,2,\ldots,m;\ j=1,2,\ldots,n)$, if $a_{ij}\le a'_{ij}$, then we have

$$\sum_{i=1}^{m} y_i a_{ij} \le \sum_{i=1}^{m} y_i a'_{ij}$$

since $y_i\ge 0$ $(i=1,2,\ldots,m)$ and $\sum_{i=1}^{m} y_i=1$, where $\boldsymbol{y}=(y_1,y_2,\ldots,y_m)^{\mathrm{T}}$ is any mixed strategy for the player I as stated earlier. Hence, we obtain

$$\min_{1\le j\le n}\{\sum_{i=1}^{m} y_i a_{ij}\} \le \min_{1\le j\le n}\{\sum_{i=1}^{m} y_i a'_{ij}\},$$

which directly infers that

$$\max_{\boldsymbol{y}\in Y}\min_{1\le j\le n}\{\sum_{i=1}^{m} y_i a_{ij}\} \le \max_{\boldsymbol{y}\in Y}\min_{1\le j\le n}\{\sum_{i=1}^{m} y_i a'_{ij}\},$$

i.e., $\upsilon((a_{ij}))\le\upsilon((a'_{ij}))$ or $\upsilon(\boldsymbol{A})\le\upsilon(\boldsymbol{A}')$, where $\boldsymbol{A}'=(a'_{ij})_{m\times n}$ is the payoff matrix of the matrix game $\boldsymbol{A}'$.

For any given values a_{ij} in the interval-valued payoffs $\bar{a}_{ij}$ $(i=1,2,\ldots,m;\ j=1,2,\ldots,n)$, according to the duality theorem of linear programming, the value $\upsilon((a_{ij}))$ (or $\upsilon(\boldsymbol{A})$) is equal to $\omega((a_{ij}))$ (or $\omega(\boldsymbol{A})$). Hence, the matrix game $\boldsymbol{A}=(a_{ij})_{m\times n}$ has a value, denoted by $V=V((a_{ij}))$ or $V=V(\boldsymbol{A})$. Thus, according to the above discussion, the value $V=V((a_{ij}))$ (or $V=V(\boldsymbol{A})$) of the matrix game $\boldsymbol{A}$ is also a non-decreasing function of the values a_{ij} $(i=1,2,\ldots,m;\ j=1,2,\ldots,n)$ in the interval-valued payoffs $\bar{a}_{ij}$.

1.6.2 Auxiliary Linear Programming Models of Interval-Valued Matrix Games

Stated as earlier, the value of the interval-valued matrix game $\bar{\boldsymbol{A}}$ should be a closed interval as well. Noticing the fact that the value $v=\upsilon((a_{ij}))$ of the matrix game $\boldsymbol{A}=(a_{ij})_{m\times n}$ is a non-decreasing function of the values a_{ij} $(i=1,2,\ldots,m;$

$j = 1, 2, \ldots, n$) in the interval-valued payoffs $\bar{a}_{ij}$. Hence, the upper bound v_R of the interval-type value of the interval-valued matrix game $\bar{\boldsymbol{A}}$ and corresponding optimal (mixed) strategy $\boldsymbol{y}_R^* \in Y$ for the player I are $v_R = \upsilon((a_{Rij}))$ and $\boldsymbol{y}_R^* = \boldsymbol{y}^*((a_{Rij}))$, respectively. According to Eq. (1.5), $(v_R, \boldsymbol{y}_R^*)$ is an optimal solution to the linear programming model as follows:

$$\max\{\upsilon_R\}$$
$$\text{s.t.}\begin{cases} \sum_{i=1}^{m} a_{Rij} y_{Ri} \geq \upsilon_R \quad (j = 1, 2, \ldots, n) \\ \sum_{i=1}^{m} y_{Ri} = 1 \\ y_{Ri} \geq 0 \quad (i = 1, 2, \ldots, m) \\ \upsilon_R \text{ unrestricted in sign,} \end{cases} \tag{1.36}$$

where y_{Ri} $(i = 1, 2, \ldots, m)$ and υ_R are variables.

To solve Eq. (1.36), let

$$x_{Ri} = \frac{y_{Ri}}{\upsilon_R} \quad (i = 1, 2, \ldots, m). \tag{1.37}$$

Without loss of generality, according to Theorem 1.2, assume that $\upsilon_R > 0$. Then, $x_{Ri} \geq 0$ $(i = 1, 2, \ldots, m)$ and

$$\sum_{i=1}^{m} x_{Ri} = \sum_{i=1}^{m} \frac{y_{Ri}}{\upsilon_R} = \frac{1}{\upsilon_R}. \tag{1.38}$$

Hence, Eq. (1.36) can be transformed into the linear programming model as follows:

$$\min\{\sum_{i=1}^{m} x_{Ri}\}$$
$$\text{s.t.}\begin{cases} \sum_{i=1}^{m} a_{Rij} x_{Ri} \geq 1 \quad (j = 1, 2, \ldots, n) \\ x_{Ri} \geq 0 \quad (i = 1, 2, \ldots, m), \end{cases} \tag{1.39}$$

where x_{Ri} $(i = 1, 2, \ldots, m)$ are variables.

Solving Eq. (1.39) through using the simplex method for linear programming, we easily obtain its optimal solution, denoted by $\boldsymbol{x}_R^* = (x_{R1}^*, x_{R2}^*, \ldots, x_{Rm}^*)^{\mathrm{T}}$. Thus, according to Eqs. (1.37) and (1.38), the upper bound v_R of the interval-type value of the interval-valued matrix game $\bar{\boldsymbol{A}}$ and corresponding optimal (optimal) strategy $\boldsymbol{y}_R^* \in Y$ for the player I are obtained as follows:

$$v_R = \frac{1}{\sum_{i=1}^{m} x_{Ri}^*} \tag{1.40}$$

and

$$y_{Ri}^* = v_R x_{Ri}^* \quad (i = 1, 2, \ldots, m), \tag{1.41}$$

respectively.

Similarly, the lower bound v_L of the interval-type value of the interval-valued matrix game $\bar{\boldsymbol{A}}$ and corresponding optimal (optimal) strategy $\boldsymbol{y}_L^* \in Y$ for the player I are $v_L = \upsilon((a_{Lij}))$ and $\boldsymbol{y}_L^* = \boldsymbol{y}^*((a_{Lij}))$, respectively. According to Eq. (1.5), $(v_L, \boldsymbol{y}_L^*)$ is an optimal solution to the linear programming model as follows:

$$\max\{\upsilon_L\}$$
$$\text{s.t.} \begin{cases} \sum_{i=1}^{m} a_{Lij} y_{Li} \geq \upsilon_L \quad (j = 1, 2, \ldots, n) \\ \sum_{i=1}^{m} y_{Li} = 1 \\ y_{Li} \geq 0 \quad (i = 1, 2, \ldots, m) \\ \upsilon_L \text{ unrestricted in sign,} \end{cases} \tag{1.42}$$

where y_{Li} $(i = 1, 2, \ldots, m)$ and υ_L are decision variables.

Let

$$x_{Li} = \frac{y_{Li}}{\upsilon_L} \quad (i = 1, 2, \ldots, m). \tag{1.43}$$

According to Theorem 1.2, without loss of generality, assume that $\upsilon_L > 0$. Then, $x_{Li} \geq 0$ $(i = 1, 2, \ldots, m)$ and

$$\sum_{i=1}^{m} x_{Li} = \sum_{i=1}^{m} \frac{y_{Li}}{\upsilon_L} = \frac{1}{\upsilon_L}. \tag{1.44}$$

Then, Eq. (1.42) can be transformed into the linear programming model as follows:

$$\min\{\sum_{i=1}^{m} x_{Li}\}$$
$$\text{s.t.} \begin{cases} \sum_{i=1}^{m} a_{Lij} x_{Li} \geq 1 \quad (j = 1, 2, \ldots, n) \\ x_{Li} \geq 0 \quad (i = 1, 2, \ldots, m), \end{cases} \tag{1.45}$$

where x_{Li} $(i = 1, 2, \ldots, m)$ are decision variables.

Solving Eq. (1.45) by using the simplex method of linear programming, we easily obtain its optimal solution, denoted by $\boldsymbol{x}_L^* = (x_{L1}^*, x_{L2}^*, \ldots, x_{Lm}^*)^{\mathrm{T}}$ $(i = 1, 2, \ldots, m)$. Thus, according to Eqs. (1.43) and (1.44), the lower bound v_L of the interval-type value of the interval-valued matrix game $\bar{\boldsymbol{A}}$ and corresponding optimal (mixed) strategy $\boldsymbol{y}_L^* \in Y$ for the player I are obtained as follows:

$$v_L = \frac{1}{\sum_{i=1}^{m} x_{Li}^*} \tag{1.46}$$

and

$$y_{Li}^* = v_L x_{Li}^* \quad (i = 1, 2, \ldots, m), \tag{1.47}$$

respectively.

Thus, we can obtain the lower bound v_L and the upper bound v_R of the interval-type value of the interval-valued matrix game $\bar{\boldsymbol{A}}$ for the player I. Therefore, the value of the interval-valued matrix game $\bar{\boldsymbol{A}}$ for the player I can be obtained as a closed interval $\bar{v} = [v_L, v_R]$.

In the same analysis to that of the player I, the upper bound μ_R of the interval-type value of the interval-valued matrix game $\bar{\boldsymbol{A}}$ and corresponding optimal (mixed) strategy $\boldsymbol{z}_R^* \in Z$ for the player II are $\mu_R = \omega((a_{Rij}))$ and $\boldsymbol{z}_R^* = \boldsymbol{z}^*((a_{Rij}))$, respectively. According to Eq. (1.8), $(\mu_R, \boldsymbol{z}_R^*)$ is an optimal solution to the linear programming model as follows:

$$\min\{\omega_R\}$$
$$\text{s.t.} \begin{cases} \sum_{j=1}^{n} a_{Rij} z_{Rj} \leq \omega_R \quad (i = 1, 2, \ldots, m) \\ \sum_{j=1}^{n} z_{Rj} = 1 \\ z_{Rj} \geq 0 \quad (j = 1, 2, \ldots, n) \\ \omega_R \text{ unrestricted in sign,} \end{cases} \tag{1.48}$$

where z_{Rj} $(j = 1, 2, \ldots, n)$ and ω_R are decision variables.

To solve Eq. (1.48), let

$$t_{Rj} = \frac{z_{Rj}}{\omega_R} \quad (j = 1, 2, \ldots, n). \tag{1.49}$$

According to Theorem 1.2, without loss of generality, assume that $\omega_R > 0$. Then, $t_{Rj} \geq 0$ $(j = 1, 2, \ldots, n)$ and

$$\sum_{j=1}^{n} t_{Rj} = \sum_{j=1}^{n} \frac{z_{Rj}}{\omega_R} = \frac{1}{\omega_R}. \tag{1.50}$$

Hence, Eq. (1.50) can be transformed into the linear programming model as follows:

$$\max\{\sum_{j=1}^{n} t_{Rj}\}$$
$$\text{s.t.}\begin{cases} \sum_{j=1}^{n} a_{Rij}t_{Rj} \leq 1 \quad (i = 1, 2, \ldots, m) \\ t_{Rj} \geq 0 \quad (j = 1, 2, \ldots, n), \end{cases} \tag{1.51}$$

where t_{Rj} $(j = 1, 2, \ldots, n)$ are variables.

Solving Eq. (1.51) by using the simplex method of linear programming, we can obtain its optimal solution, denoted by $\boldsymbol{t}_R^* = (t_{R1}^*, t_{R2}^*, \ldots, t_{Rn}^*)^{\mathrm{T}}$. Therefore, according to Eqs. (1.49) and (1.50), the upper bound μ_R of the interval-type value of the interval-valued matrix game $\bar{\boldsymbol{A}}$ and corresponding optimal (mixed) strategy $\boldsymbol{z}_R^* \in Z$ for the player II are obtained as follows:

$$\mu_R = \frac{1}{\sum_{j=1}^{n} t_{Rj}^*} \tag{1.52}$$

and

$$z_{Rj}^* = \mu_R t_{Rj}^* \quad (j = 1, 2, \ldots, n), \tag{1.53}$$

respectively.

Analogously, the lower bound μ_L of the interval-type value of the interval-valued matrix game $\bar{\boldsymbol{A}}$ and corresponding optimal (mixed) strategy $\boldsymbol{z}_L^* \in Z$ for the player II are $\mu_L = \omega((a_{Lij}))$ and $\boldsymbol{z}_L^* = \boldsymbol{z}^*((a_{Lij}))$, respectively. According to Eq. (1.8), $(\mu_L, \boldsymbol{z}_L^*)$ is an optimal solution to the linear programming model as follows:

$$\min\{\omega_L\}$$
$$\text{s.t.}\begin{cases} \sum_{j=1}^{n} a_{Lij}z_{Lj} \leq \omega_L \quad (i = 1, 2, \ldots, m) \\ \sum_{j=1}^{n} z_{Lj} = 1 \\ z_{Lj} \geq 0 \quad (j = 1, 2, \ldots, n) \\ \omega_L \text{ unrestricted in sign,} \end{cases} \tag{1.54}$$

where z_{Lj} $(j = 1, 2, \ldots, n)$ and ω_L are decision variables.

Let

$$t_{Lj} = \frac{z_{Lj}}{\omega_L} \quad (j = 1, 2, \ldots, n). \tag{1.55}$$

Without loss of generality, assume that $\omega_L > 0$. Then, $t_{Lj} \geq 0$ $(j = 1, 2, \ldots, n)$ and

$$\sum_{j=1}^{n} t_{Lj} = \sum_{j=1}^{n} \frac{z_{Lj}}{\omega_L} = \frac{1}{\omega_L}. \tag{1.56}$$

Hence, Eq. (1.54) can be transformed into the linear programming model as follows:

$$\max\{\sum_{j=1}^{n} t_{Lj}\}$$
$$\text{s.t.} \begin{cases} \sum_{j=1}^{n} a_{Lij} t_{Lj} \leq 1 \quad (i = 1, 2, \ldots, m) \\ t_{Lj} \geq 0 \quad (j = 1, 2, \ldots, n), \end{cases} \tag{1.57}$$

where t_{Lj} $(j = 1, 2, \ldots, n)$ are variables.

Solving Eq. (1.57) through using the simplex method of linear programming, we easily obtain its optimal solution, denoted by $\boldsymbol{t}_L^* = (t_{L1}^*, t_{L2}^*, \ldots, t_{Ln}^*)^{\mathrm{T}}$. According to Eqs. (1.55) and (1.56), the lower bound μ_L of the interval-type value of the interval-valued matrix game $\bar{\boldsymbol{A}}$ and corresponding optimal (mixed) strategy $\boldsymbol{z}_L^* \in Z$ for the player II are obtained as follows:

$$\mu_L = \frac{1}{\sum_{j=1}^{n} t_{Lj}^*} \tag{1.58}$$

and

$$z_{Lj}^* = \mu_L t_{Lj}^* \quad (j = 1, 2, \ldots, n), \tag{1.59}$$

respectively.

It is easy to see that Eqs. (1.39) and (1.51) are a pair of primal-dual linear programming models. Therefore, the minimum of $\sum_{i=1}^{m} x_{Ri}$ (i.e., the maximum of υ_R) is equal to the maximum of $\sum_{j=1}^{n} t_{Rj}$ (i.e., the minimum of ω_R) by the duality theorem of linear programming, i.e., $v_R = \mu_R$. Similarly, Eqs. (1.45) and (1.57) are a pair of primal-dual linear programming models. Hence, $v_L = \mu_L$. Thus, the players I and II have the identical interval-type value of the interval-valued matrix game. Namely, the value of the interval-valued matrix game $\bar{\boldsymbol{A}}$ is a closed interval $\bar{V} = [V_L, V_R]$, where $V_L = v_L = \mu_L$ and $V_R = v_R = \mu_R$.

From the aforementioned discussion, the value $\bar{V} = [V_L, V_R]$ of the interval-valued matrix game $\bar{A}$ can be obtained through solving either Eqs. (1.39) and (1.45) or Eqs. (1.51) and (1.57) by directly using the simplex method of linear programming.

1.6.3 Real Example Analysis of Investment Decision Problems

There are many applications of the classical game theory to real decision problems in finance, management, business, and economics. In particular, the following is an example how interval-valued matrix games are applied to determine optimal investment strategies [12].

(1) Investment decision problems and a real numerical example

Let us consider the case of an investor (i.e., the player I) making a decision as to how to invest a non-divisible sum of money when the economic environment (i.e., the player II) may be categorized into a finite number of states. There is no guarantee that any single value (return on the investment) can adequately model the payoff for any one of the economic states. Hence, it is more realistic and appropriate to assume that each payoff belongs to some interval.

For this example, it is assumed that the decision of such an investor (i.e., the player I) can be modeled under the assumption that the economic environment/nature (i.e., the player II) is, in fact, a rational "player" that will choose an optimal strategy. Suppose that the options (i.e., pure strategies) for this player II are: strong economic growth (β_1), moderate economic growth (β_2), no growth nor shrinkage (β_3), and moderate shrinkage or negative growth (β_4). For the investor, the options (i.e., pure strategies) are: invest in bonds (δ_1), invest in stocks (δ_2), and invest in a guaranteed fixed return account (δ_3). In this case, clearly a single value for the payoff of either investment in bonds or stocks cannot be realistically modeled by an exact value representing the percent of the return. Therefore, an interval-valued matrix game can suitably represent the view of the game from both the players' perspectives.

Then, let us consider the following specific interval-valued matrix $\bar{A}_0$ for this scenario, where the percentage of the return represented in decimal form for each outcome is given in the interval format as follows:

$$\bar{A}_0 = \begin{array}{c} \\ \delta_1 \\ \delta_2 \\ \delta_3 \end{array} \begin{array}{c} \begin{array}{cccc} \quad\beta_1 \quad & \quad\beta_2\quad & \quad\beta_3\quad & \quad\beta_4\quad \end{array} \\ \begin{pmatrix} [0.12, 0.17] & [0.11, 0.16] & [0.075, 0.12] & [0.068, 0.13] \\ [0.18, 0.22] & [0.12, 0.15] & [0.072, 0.14] & [-0.05, 0.15] \\ [0.043, 0.043] & [0.043, 0.043] & [0.043, 0.043] & [0.043, 0.043] \end{pmatrix} \end{array},$$

where the interval [0.12, 0.17] means that the percentage of the return for the investor (i.e., player I) is between 12 and 17 % if he/she invests in bonds (i.e., chooses the pure strategy δ_1) when the economic environment/nature is strong growth (i.e., the player II chooses the pure strategy β_1). Other entries (i.e., the intervals) in the interval-valued payoff matrix $\bar{\boldsymbol{A}}_0$ can be similarly explained.

Now, the investor tries to determine the range of the percentage of the expected return in decimal form. Namely, the lower and upper bounds of the interval-type value of the interval-valued matrix game $\bar{\boldsymbol{A}}_0$ need to be determined.

(2) Computational results obtained by different methods and analysis

In this subsection, the aforementioned numerical example is solved by the linear programming method proposed in the above Sect. 1.6.2 and other similar methods [18, 19, 28, 29]. The computational results are analyzed and compared to show the validity, applicability, and superiority of the developed linear programming method.

(2a) Computational results obtained by the developed linear programming method

According to the linear programming method proposed in the above Sect. 1.6.2, i.e., using Eqs. (1.39) and (1.45), the linear programming models are constructed as follows:

$$\min\{x_{R1}+x_{R2}+x_{R3}\}$$
$$\text{s.t.}\begin{cases} 0.17x_{R1}+0.22x_{R2}+0.043x_{R3}\geq 1\\ 0.16x_{R1}+0.15x_{R2}+0.043x_{R3}\geq 1\\ 0.12x_{R1}+0.14x_{R2}+0.043x_{R3}\geq 1\\ 0.13x_{R1}+0.15x_{R2}+0.043x_{R3}\geq 1\\ x_{Ri}\geq 0\quad (i=1,2,3)\end{cases} \tag{1.60}$$

and

$$\min\{x_{L1}+x_{L2}+x_{L3}\}$$
$$\text{s.t.}\begin{cases} 0.12x_{L1}+0.18x_{L2}+0.043x_{L3}\geq 1\\ 0.11x_{L1}+0.12x_{L2}+0.043x_{L3}\geq 1\\ 0.075x_{L1}+0.072x_{L2}+0.043x_{L3}\geq 1\\ 0.068x_{L1}-0.05x_{L2}+0.043x_{L3}\geq 1\\ x_{Li}\geq 0\quad (i=1,2,3),\end{cases} \tag{1.61}$$

respectively.

Solving Eqs. (1.60) and (1.61) by using the simplex method of linear programming, we can easily obtain their optimal solutions as follows:

$$\boldsymbol{x}_R^*=(x_{R1}^*,x_{R2}^*,x_{R3}^*)^{\mathrm{T}}=(0,7.143,0)^{\mathrm{T}}$$

and

$$\boldsymbol{x}_L^* = (x_{L1}^*, x_{L2}^*, x_{L3}^*)^{\mathrm{T}} = (14.706, 0, 0)^{\mathrm{T}},$$

respectively. According to Eqs. (1.40), (1.41), (1.46), and (1.47), we have

$$v_R = 0.14, y_{R1}^* = 0, y_{R2}^* = 1, y_{R3}^* = 0$$

and

$$v_L = 0.068, y_{L1}^* = 1, y_{L2}^* = 0, y_{L3}^* = 0.$$

Thus, the value of the interval-valued matrix game $\bar{\boldsymbol{A}}_0$ for the investor is a closed interval $\bar{v} = [v_L, v_R] = [0.068, 0.14]$.

Analogously, using Eqs. (1.51) and (1.57), the linear programming models can be constructed as follows:

$$\max\{t_{R1} + t_{R2} + t_{R3} + t_{R4}\}$$
$$\text{s.t.} \begin{cases} 0.17t_{R1} + 0.16t_{R2} + 0.12t_{R3} + 0.13t_{R4} \leq 1 \\ 0.22t_{R1} + 0.15t_{R2} + 0.14t_{R3} + 0.15t_{R4} \leq 1 \\ 0.043t_{R1} + 0.043t_{R2} + 0.043t_{R3} + 0.043t_{R4} \leq 1 \\ t_{Rj} \geq 0 \quad (j = 1, 2, 3, 4) \end{cases} \tag{1.62}$$

and

$$\max\{t_{L1} + t_{L2} + t_{L3} + t_{L4}\}$$
$$\text{s.t.} \begin{cases} 0.12t_{L1} + 0.11t_{L2} + 0.075t_{L3} + 0.068t_{L4} \leq 1 \\ 0.18t_{L1} + 0.12t_{L2} + 0.072t_{L3} - 0.05t_{L4} \leq 1 \\ 0.043t_{L1} + 0.043t_{L2} + 0.043t_{L3} + 0.043t_{L4} \leq 1 \\ t_{Lj} \geq 0 \quad (j = 1, 2, 3, 4), \end{cases} \tag{1.63}$$

respectively.

Solving Eqs. (1.62) and (1.63) by using the simplex method of linear programming, we can easily obtain their optimal solutions as follows:

$$\boldsymbol{t}_R^* = (t_{R1}^*, t_{R2}^*, t_{R3}^*, t_{R4}^*)^{\mathrm{T}} = (0, 0, 7.143, 0)^{\mathrm{T}}$$

and

$$\boldsymbol{t}_L^* = (t_{L1}^*, t_{L2}^*, t_{L3}^*, t_{L4}^*)^{\mathrm{T}} = (0, 0, 0, 14.706)^{\mathrm{T}},$$

respectively. According to Eqs. (1.52), (1.53), (1.58) and (1.59), we have

$$\mu_R = 0.14, z_{R1}^* = 0, z_{R2}^* = 0, z_{R3}^* = 1, z_{R4}^* = 0$$

and

$$\mu_L = 0.068, z_{L1}^* = 0, z_{L2}^* = 0, z_{L3}^* = 0, z_{L4}^* = 1.$$

Hence, the value of the interval-valued matrix game $\bar{A}_0$ for the economic environment/nature (i.e., player II) is a closed interval $\bar{\mu} = [\mu_L, \mu_R] = [0.068, 0.14]$.

Obviously, $\bar{v} = \bar{\mu} = [0.068, 0.14]$, i.e., the investor and the economic environment/nature have the identical interval-type value of the interval-valued matrix game. Therefore, the value of the interval-valued matrix game $\bar{A}_0$ is obtained as an interval $\bar{V} = \bar{v} = \bar{\mu} = [0.068, 0.14]$, which means that the percentage of the expected return for the investor is between 6.8 and 14 %. In other words, the investor's minimum percentage of the expected return is 6.8 % while his/her maximum percentage of the expected return is 14 %. He/she could obtain any intermediate value (%) in the interval $\bar{V} = [0.068, 0.14]$ (i.e., between 6.8 and 14 %) as the percentage of the expected return.

(2b) Computational results obtained by Liu and Kao' method

According to the method [i.e., Eq. (7)] given by Liu and Kao [18], the upper bound $\bar{v}^{\mathrm{LK}}$ of the interval-type value of the interval-valued matrix game $\bar{A}_0$ in this example for the investor (i.e., player I) can be obtained through solving the linear programming model as follows:

$$\max\{\bar{\upsilon}^{\mathrm{LK}}\}$$
$$\text{s.t.}\begin{cases} p_{11} + p_{21} + p_{31} \geq \bar{\upsilon}^{\mathrm{LK}} \\ p_{12} + p_{22} + p_{32} \geq \bar{\upsilon}^{\mathrm{LK}} \\ p_{13} + p_{23} + p_{33} \geq \bar{\upsilon}^{\mathrm{LK}} \\ p_{14} + p_{24} + p_{34} \geq \bar{\upsilon}^{\mathrm{LK}} \\ 0.12y_1 \leq p_{11} \leq 0.17y_1 \\ 0.11y_1 \leq p_{12} \leq 0.16y_1 \\ 0.075y_1 \leq p_{13} \leq 0.12y_1 \\ 0.068y_1 \leq p_{14} \leq 0.13y_1 \\ 0.18y_2 \leq p_{21} \leq 0.22y_2 \\ 0.12y_2 \leq p_{22} \leq 0.15y_2 \\ 0.072y_2 \leq p_{23} \leq 0.14y_2 \\ -0.05y_2 \leq p_{24} \leq 0.15y_2 \\ p_{31} = 0.043y_3 \\ p_{32} = 0.043y_3 \\ p_{33} = 0.043y_3 \\ p_{34} = 0.043y_3 \\ y_1 + y_2 + y_3 = 1 \\ y_i \geq 0 \quad (i = 1, 2, 3) \\ \bar{\upsilon}^{\mathrm{LK}} \text{ and } p_{ij} \text{ unrestricted in sign } (i = 1, 2, 3;\ j = 1, 2, 3, 4), \end{cases}$$

where y_i, p_{ij} $(i = 1, 2, 3; j = 1, 2, 3, 4)$, and $\bar{\upsilon}^{\mathrm{LK}}$ are decision variables. Solving the above linear programming model by the simplex method of linear programming, we can obtain its optimal solution whose components are given as follows:

$$\begin{aligned}\bar{v}^{\mathrm{LK}} &= \max\{\bar{\upsilon}^{\mathrm{LK}}\} = 0.14, y_1^* = 0, y_2^* = 1, y_3^* = 0, p_{11}^* = 0, p_{12}^* \\ &= 0, p_{13}^* = 0, p_{14}^* = 0, p_{21}^* = 0.18, p_{22}^* = 0.14, p_{23}^* \\ &= 0.14, p_{24}^* = 0.14, p_{31}^* = 0, p_{32}^* = 0, p_{33}^* = 0, p_{34}^* = 0.\end{aligned}$$

In the same way, using Eq. (1.9) given by Liu and Kao [18], the lower bound $\underline{v}^{\mathrm{LK}}$ of the interval-type value of the interval-valued matrix game $\bar{\boldsymbol{A}}_0$ for the investor can be obtained through solving the linear programming model as follows:

$$\min\{\underline{\upsilon}^{\mathrm{LK}}\}$$

$$\text{s.t.}\begin{cases} q_{11} + q_{12} + q_{13} + q_{14} \leq \underline{\upsilon}^{\mathrm{LK}} \\ q_{21} + q_{22} + q_{23} + q_{24} \leq \underline{\upsilon}^{\mathrm{LK}} \\ q_{31} + q_{32} + q_{33} + q_{34} \leq \underline{\upsilon}^{\mathrm{LK}} \\ 0.12z_1 \leq q_{11} \leq 0.17z_1 \\ 0.18z_1 \leq q_{21} \leq 0.22z_1 \\ q_{31} = 0.043z_1 \\ 0.11z_2 \leq q_{12} \leq 0.16z_2 \\ 0.12z_2 \leq q_{22} \leq 0.15z_2 \\ q_{32} = 0.043z_2 \\ 0.075z_3 \leq q_{13} \leq 0.12z_3 \\ 0.072z_3 \leq q_{23} \leq 0.14z_3 \\ q_{33} = 0.043z_3 \\ 0.068z_4 \leq q_{14} \leq 0.13z_4 \\ -0.05z_4 \leq q_{24} \leq 0.15z_4 \\ q_{34} = 0.043z_4 \\ z_1 + z_2 + z_3 + z_4 = 1 \\ z_j \geq 0 \quad (j = 1, 2, 3, 4) \\ \underline{\upsilon}^{\mathrm{LK}} \text{ and } q_{ij} \text{ unrestricted in sign } (i = 1, 2, 3;\ j = 1, 2, 3, 4), \end{cases}$$

where z_j, q_{ij} $(i = 1, 2, 3;\ j = 1, 2, 3, 4)$, and $\underline{\upsilon}^{\mathrm{LK}}$ are decision variables. Solving the above linear programming model by the simplex method of linear programming, we can obtain its optimal solution whose components are given as follows:

$$\underline{v}^{\mathrm{LK}} = \max\{\underline{\upsilon}^{\mathrm{LK}}\} = 0.14, z_1^* = 0, z_2^* = 0, z_3^* = 0, z_4^* = 1, q_{11}^* \\ = 0, q_{12}^* = 0, q_{13}^* = 0, q_{14}^* = 0.068, q_{21}^* = 0, q_{22}^* = 0, \\ q_{23}^* = 0, q_{24}^* = 0, q_{31}^* = 0, q_{32}^* = 0, q_{33}^* = 0, q_{34}^* = 0.$$

Thus, it directly follows that the value of the interval-valued matrix game $\bar{\boldsymbol{A}}_0$ for the investor is a closed interval $[\underline{v}^{\mathrm{LK}}, \bar{v}^{\mathrm{LK}}] = [0.068, 0.14]$.

Analogously, we can construct two linear programming models for determining the lower bound $\underline{\mu}^{\mathrm{LK}}$ and the upper bound $\bar{\mu}^{\mathrm{LK}}$ of the interval-type value of the interval-valued matrix game $\bar{\boldsymbol{A}}_0$ for the economic environment/nature (i.e., player II), which are computed as $\underline{\mu}^{\mathrm{LK}} = 0.068$ and $\bar{\mu}^{\mathrm{LK}} = 0.14$. Namely, the value of the interval-valued matrix game $\bar{\boldsymbol{A}}_0$ for the player II is a closed interval $[\underline{\mu}^{\mathrm{LK}}, \bar{\mu}^{\mathrm{LK}}] = [0.068, 0.14]$. Thus, the players I and II have the identical interval-type value of the interval-valued matrix game. Hence, the value of the interval-valued matrix game $\bar{\boldsymbol{A}}_0$ is a closed interval $[\underline{V}^{\mathrm{LK}}, \bar{V}^{\mathrm{LK}}] = [0.068, 0.14]$.

(2c) Computational results obtained by Li's model

Li [29] and Li and Yang [35] developed the two-level linear programming method of fuzzy matrix games with payoffs of triangular fuzzy numbers, which was called as Li's model by Bector and Chandra [4] and Larbani [17]. In Li's model, assume that the value of a fuzzy matrix game with payoffs of triangular fuzzy numbers is also a triangular fuzzy number, which has three parameters including the mean and the lower and upper bounds/limits of the triangular fuzzy number.

Stated as earlier, from the viewpoint of logic, the player I's gain-floor and the player II's loss-ceiling in the interval-valued matrix game $\bar{\boldsymbol{A}}_0$ in the above example should be intervals, denoted by $[\underline{\upsilon}^{\mathrm{L}}, \bar{\upsilon}^{\mathrm{L}}]$ and $[\underline{\omega}^{\mathrm{L}}, \bar{\omega}^{\mathrm{L}}]$, respectively. To employ Li's model to solve the above numerical example, $\upsilon^{\mathrm{L}} = (\underline{\upsilon}^{\mathrm{L}} + \bar{\upsilon}^{\mathrm{L}})/2$ and $\omega^{\mathrm{L}} = (\underline{\omega}^{\mathrm{L}} + \bar{\omega}^{\mathrm{L}})/2$ are taken as the means of the intervals $[\underline{\upsilon}^{\mathrm{L}}, \bar{\upsilon}^{\mathrm{L}}]$ and $[\underline{\omega}^{\mathrm{L}}, \bar{\omega}^{\mathrm{L}}]$, respectively. Then, the linear programming model in the level 1 for the investor (i.e., player I) is constructed as follows:

$$
\max\{\upsilon^{\mathrm{L}}\}
$$

$$
\text{s.t.}\begin{cases}
0.12y_1+0.18y_2+0.043y_3 \geq \underline{\upsilon}^{\mathrm{L}} \\
0.11y_1+0.12y_2+0.043y_3 \geq \underline{\upsilon}^{\mathrm{L}} \\
0.075y_1+0.072y_2+0.043y_3 \geq \underline{\upsilon}^{\mathrm{L}} \\
0.068y_1-0.05y_2+0.043y_3 \geq \underline{\upsilon}^{\mathrm{L}} \\
0.145y_1+0.2y_2+0.043y_3 \geq \upsilon^{\mathrm{L}} \\
0.135y_1+0.135y_2+0.043y_3 \geq \upsilon^{\mathrm{L}} \\
0.0975y_1+0.106y_2+0.043y_3 \geq \upsilon^{\mathrm{L}} \\
0.099y_1+0.05y_2+0.043y_3 \geq \upsilon^{\mathrm{L}} \\
0.17y_1+0.22y_2+0.043y_3 \geq \bar{\upsilon}^{\mathrm{L}} \\
0.16y_1+0.15y_2+0.043y_3 \geq \bar{\upsilon}^{\mathrm{L}} \\
0.12y_1+0.14y_2+0.043y_3 \geq \bar{\upsilon}^{\mathrm{L}} \\
0.13y_1+0.15y_2+0.043y_3 \geq \bar{\upsilon}^{\mathrm{L}} \\
y_1+y_2+y_3=1 \\
y_i \geq 0 \quad (i=1,2,3) \\
\bar{\upsilon}^{\mathrm{L}} \geq \upsilon^{\mathrm{L}} \geq \underline{\upsilon}^{\mathrm{L}} \\
\bar{\upsilon}^{\mathrm{L}}, \upsilon^{\mathrm{L}}, \text{ and } \underline{\upsilon}^{\mathrm{L}} \text{ unrestricted in sign,}
\end{cases} \tag{1.64}
$$

where $y_i\ (i=1,2,3)$, $\bar{\upsilon}^{\mathrm{L}}, \upsilon^{\mathrm{L}}$, and $\underline{\upsilon}^{\mathrm{L}}$ are decision variables.

Solving Eq. (1.64) by using the simplex method of linear programming, we obtain its optimal solution whose components are given as follows:

$$
v^{\mathrm{L}}=\max\{\upsilon^{\mathrm{L}}\}=0.094, y_1^*=1, y_2^*=0, y_3^*=0, \underline{\upsilon}_0^L=0.12, \underline{\upsilon}_0^L=0.068.
$$

Hereby, according to Li's model [29, 35], two linear programming models in the level 2 for the investor are constructed as follows:

$$
\max\{\underline{\upsilon}^{\mathrm{L}}\}
$$

$$
\text{s.t.}\begin{cases}
0.12y_1^*+0.18y_2^*+0.043y_3^* \geq \underline{\upsilon}^{\mathrm{L}} \\
0.11y_1^*+0.12y_2^*+0.043y_3^* \geq \underline{\upsilon}^{\mathrm{L}} \\
0.075y_1^*+0.072y_2^*+0.043y_3^* \geq \underline{\upsilon}^{\mathrm{L}} \\
0.068y_1^*-0.05y_2^*+0.043y_3^* \geq \underline{\upsilon}^{\mathrm{L}} \\
\underline{\upsilon}^{\mathrm{L}} \geq 0.068 \\
\underline{\upsilon}^{\mathrm{L}} \text{ unrestricted in sign}
\end{cases} \tag{1.65}
$$

and

$$\max\{\bar{\upsilon}^{\mathrm{L}}\}$$
$$\text{s.t.}\begin{cases} 0.17y_1^* + 0.22y_2^* + 0.043y_3^* \geq \bar{\upsilon}^{\mathrm{L}} \\ 0.16y_1^* + 0.15y_2^* + 0.043y_3^* \geq \bar{\upsilon}^{\mathrm{L}} \\ 0.12y_1^* + 0.14y_2^* + 0.043y_3^* \geq \bar{\upsilon}^{\mathrm{L}} \\ 0.13y_1^* + 0.15y_2^* + 0.043y_3^* \geq \bar{\upsilon}^{\mathrm{L}} \\ \bar{\upsilon}^{\mathrm{L}} \geq 0.12 \\ \bar{\upsilon}^{\mathrm{L}} \text{ unrestricted in sign,} \end{cases} \tag{1.66}$$

where $\bar{\upsilon}^{\mathrm{L}}$ and $\underline{\upsilon}^{\mathrm{L}}$ are decision variables.

Solving Eqs. (1.65) and (1.66) by applying the simplex method of linear programming, we can obtain their optimal solutions as follows:

$$\underline{v}^{\mathrm{L}} = \max\{\underline{\upsilon}^{\mathrm{L}}\} = 0.068$$

and

$$\bar{v}^{\mathrm{L}} = \max\{\bar{\upsilon}^{\mathrm{L}}\} = 0.12,$$

respectively. Therefore, the value of the interval-valued matrix game $\bar{\boldsymbol{A}}_0$ for the investor is a closed interval $[\underline{v}^{\mathrm{L}}, \bar{v}^{\mathrm{L}}] = [0.068, 0.12]$.

Analogously, the linear programming model in the level 1 for the economic environment/nature (i.e., player II) is constructed as follows:

$$\min\{\omega^{\mathrm{L}}\}$$
$$\text{s.t.}\begin{cases} 0.12z_1 + 0.11z_2 + 0.075z_3 + 0.068z_4 \leq \underline{\omega}^{\mathrm{L}} \\ 0.18z_1 + 0.12z_2 + 0.072z_3 - 0.05z_4 \leq \underline{\omega}^{\mathrm{L}} \\ 0.043z_1 + 0.043z_2 + 0.043z_3 + 0.043z_4 \leq \underline{\omega}^{\mathrm{L}} \\ 0.145z_1 + 0.135z_2 + 0.0975z_3 + 0.099z_4 \leq \omega^{\mathrm{L}} \\ 0.2z_1 + 0.135z_2 + 0.106z_3 + 0.05z_4 \leq \omega^{\mathrm{L}} \\ 0.043z_1 + 0.043z_2 + 0.043z_3 + 0.043z_4 \leq \omega^{\mathrm{L}} \\ 0.17z_1 + 0.16z_2 + 0.12z_3 + 0.13z_4 \leq \bar{\omega}^{\mathrm{L}} \\ 0.22z_1 + 0.15z_2 + 0.14z_3 + 0.15z_4 \leq \bar{\omega}^{\mathrm{L}} \\ 0.043z_1 + 0.043z_2 + 0.043z_3 + 0.043z_4 \leq \bar{\omega}^{\mathrm{L}} \\ z_1 + z_2 + z_3 + z_4 = 1 \\ z_j \geq 0 \quad (j = 1, 2, 3, 4) \\ \bar{\omega}^{\mathrm{L}} \geq \omega^{\mathrm{L}} \geq \underline{\omega}^{\mathrm{L}} \\ \bar{\omega}^{\mathrm{L}},\ \omega^{\mathrm{L}},\ \text{and } \underline{\omega}^{\mathrm{L}} \text{ unrestricted in sign,} \end{cases} \tag{1.67}$$

where $z_j\ (j = 1, 2, 3, 4)$, $\bar{\omega}^{\mathrm{L}}, \omega^{\mathrm{L}}$, and $\underline{\omega}^{\mathrm{L}}$ are decision variables.

Solving Eq. (1.67) by using the simplex method of linear programming, we can easily obtain its optimal solution whose components are given as follows:

$$\mu^{\mathrm{L}} = \min\{\omega^{\mathrm{L}}\} = 0.108, z_1^* = 0, z_2^* = 0, z_3^* = 1, z_4^* = 0, \bar{\omega}_0^L = 0.14, \underline{\omega}_0^L = 0.075.$$

Combining with Li's model [29, 35], two linear programming models in the level 2 for the player II (i.e., the economic environment/nature) are constructed as follows:

$$\min\{\underline{\omega}^{\mathrm{L}}\}$$
$$\text{s.t.}\begin{cases} 0.12z_1^* + 0.11z_2^* + 0.075z_3^* + 0.068z_4^* \leq \underline{\omega}^{\mathrm{L}} \\ 0.18z_1^* + 0.12z_2^* + 0.072z_3^* - 0.05z_4^* \leq \underline{\omega}^{\mathrm{L}} \\ 0.043z_1^* + 0.043z_2^* + 0.043z_3^* + 0.043z_4^* \leq \underline{\omega}^{\mathrm{L}} \\ \underline{\omega}^{\mathrm{L}} \leq 0.075 \\ \underline{\omega}^{\mathrm{L}} \text{ unrestricted in sign} \end{cases} \tag{1.68}$$

and

$$\min\{\bar{\omega}^{\mathrm{L}}\}$$
$$\text{s.t.}\begin{cases} 0.17z_1^* + 0.16z_2^* + 0.12z_3^* + 0.13z_4^* \leq \bar{\omega}^{\mathrm{L}} \\ 0.22z_1^* + 0.15z_2^* + 0.14z_3^* + 0.15z_4^* \leq \bar{\omega}^{\mathrm{L}} \\ 0.043z_1^* + 0.043z_2^* + 0.043z_3^* + 0.043z_4^* \leq \bar{\omega}^{\mathrm{L}} \\ \bar{\omega}^{\mathrm{L}} \leq 0.14 \\ \bar{\omega}^{\mathrm{L}} \text{ unrestricted in sign,} \end{cases} \tag{1.69}$$

where $\bar{\omega}^{\mathrm{L}}$ and $\underline{\omega}^{\mathrm{L}}$ are decision variables.

Solving Eqs. (1.68) and (1.69) by using the simplex method of linear programming, we can easily obtain their optimal solutions as follows:

$$\underline{\mu}^{\mathrm{L}} = \min\{\underline{\omega}^{\mathrm{L}}\} = 0.075$$

and

$$\bar{\mu}^{\mathrm{L}} = \min\{\bar{\omega}^{\mathrm{L}}\} = 0.14,$$

respectively.

Then, the value of the interval-valued matrix game $\bar{A}_0$ for the player II is a closed interval $[\underline{\mu}^{\mathrm{L}}, \bar{\mu}^{\mathrm{L}}] = [0.075, 0.14]$, which is larger than the interval-type value $[\underline{v}^{\mathrm{L}}, \bar{v}^{\mathrm{L}}] = [0.068, 0.12]$ of the interval-valued matrix game $\bar{A}_0$ for the player I.

(2d) Computational results obtained by the weighted average method

According to Eq. (13) in the weighted average method developed by Li et al. [28], the lower and upper bounds of the interval-type gain-floor and corresponding optimal (mixed) strategy for the investor (i.e., player I) can be obtained through solving the linear programming model as follows:

$$\max\{\frac{3\underline{\upsilon}^{\mathrm{LNZ}}+\bar{\upsilon}^{\mathrm{LNZ}}}{4}\}$$
$$\text{s.t.}\begin{cases} 0.12y_1+0.18y_2+0.043y_3\geq\underline{\upsilon}^{\mathrm{LNZ}}\\ 0.11y_1+0.12y_2+0.043y_3\geq\underline{\upsilon}^{\mathrm{LNZ}}\\ 0.075y_1+0.072y_2+0.043y_3\geq\underline{\upsilon}^{\mathrm{LNZ}}\\ 0.068y_1-0.05y_2+0.043y_3\geq\underline{\upsilon}^{\mathrm{LNZ}}\\ (1-\varepsilon)(0.17y_1+0.22y_2+0.043y_3)+\varepsilon(0.12y_1+0.18y_2+0.043y_3)\geq(1-\varepsilon)\underline{\upsilon}^{\mathrm{LNZ}}+\varepsilon\bar{\upsilon}^{\mathrm{LNZ}}\\ (1-\varepsilon)(0.16y_1+0.15y_2+0.043y_3)+\varepsilon(0.11y_1+0.12y_2+0.043y_3)\geq(1-\varepsilon)\underline{\upsilon}^{\mathrm{LNZ}}+\varepsilon\bar{\upsilon}^{\mathrm{LNZ}}\\ (1-\varepsilon)(0.12y_1+0.14y_2+0.043y_3)+\varepsilon(0.075y_1+0.072y_2+0.043y_3)\geq(1-\varepsilon)\underline{\upsilon}^{\mathrm{LNZ}}+\varepsilon\bar{\upsilon}^{\mathrm{LNZ}}\\ (1-\varepsilon)(0.13y_1+0.15y_2+0.043y_3)+\varepsilon(0.068y_1-0.05y_2+0.043y_3)\geq(1-\varepsilon)\underline{\upsilon}^{\mathrm{LNZ}}+\varepsilon\bar{\upsilon}^{\mathrm{LNZ}}\\ \underline{\upsilon}^{\mathrm{LNZ}}\leq\bar{\upsilon}^{\mathrm{LNZ}}\\ y_1+y_2+y_3=1\\ y_i\geq 0\quad(i=1,2,3)\\ \bar{\upsilon}^{\mathrm{LNZ}}\text{ and }\underline{\upsilon}^{\mathrm{LNZ}}\text{ unrestricted in sign,}\end{cases}\tag{1.70}$$

where y_i $(i=1,2,3)$, $\bar{\upsilon}^{\mathrm{LNZ}}$, and $\underline{\upsilon}^{\mathrm{LNZ}}$ are decision variables, the parameter $\varepsilon\in[0,1]$ expresses the acceptance degree of the interval-valued inequality constraints which may be allowed to violate. ε is determined by the players *a priori* according to the real situations.

Taking $\varepsilon=0.5$, and solving Eq. (1.70) by using the simplex method of linear programming, we can obtain its optimal solution whose components are given as follows:

$$\underline{\upsilon}^{\mathrm{LNZ}*}=0.068,\bar{\upsilon}^{\mathrm{LNZ}*}=0.127,y_1^*=1,y_2^*=0,y_3^*=0.$$

Thus, we obtain the optimal mixed strategy $\boldsymbol{y}^*=(y_1^*,y_2^*,y_3^*)^{\mathrm{T}}=(1,0,0)^{\mathrm{T}}$ and the gain-floor $[\underline{v}^{\mathrm{LNZ}},\bar{v}^{\mathrm{LNZ}}]=[\underline{\upsilon}^{\mathrm{LNZ}*},\bar{\upsilon}^{\mathrm{LNZ}*}]=[0.068,0.127]$ for the player I.

In the same way, using Eq. (16) in the weighted average method [28], the lower and upper bounds of the interval-type loss-ceiling and corresponding optimal (mixed) strategy for the economic environment/nature (i.e., player II) can be obtained through solving the linear programming model as follows:

$$\min\{\frac{\underline{\omega}^{\mathrm{LNZ}}+3\bar{\omega}^{\mathrm{LNZ}}}{4}\}$$
$$\text{s.t.}\begin{cases} 0.17z_1+0.16z_2+0.12z_3+0.13z_4\leq\bar{\omega}^{\mathrm{L}} \\ 0.22z_1+0.15z_2+0.14z_3+0.15z_4\leq\bar{\omega}^{\mathrm{L}} \\ 0.043z_1+0.043z_2+0.043z_3+0.043z_4\leq\bar{\omega}^{\mathrm{L}} \\ (1-\varepsilon)(0.12z_1+0.11z_2+0.075z_3+0.068z_4)+\varepsilon(0.17z_1+0.16z_2+0.12z_3+0.13z_4)\leq(1-\varepsilon)\underline{\omega}^{\mathrm{LNZ}}+\varepsilon\bar{\omega}^{\mathrm{LNZ}} \\ (1-\varepsilon)(0.18z_1+0.12z_2+0.072z_3-0.05z_4)+\varepsilon(0.22z_1+0.15z_2+0.14z_3+0.15z_4)\leq(1-\varepsilon)\underline{\omega}^{\mathrm{LNZ}}+\varepsilon\bar{\omega}^{\mathrm{LNZ}} \\ (1-\varepsilon)(0.043z_1+0.043z_2+0.043z_3+0.043z_4)+\varepsilon(0.043z_1+0.043z_2+0.043z_3+0.043z_4)\leq(1-\varepsilon)\underline{\omega}^{\mathrm{LNZ}}+\varepsilon\bar{\omega}^{\mathrm{LNZ}} \\ \underline{\omega}^{\mathrm{LNZ}}\leq\bar{\omega}^{\mathrm{LNZ}} \\ z_1+z_2+z_3+z_4=1 \\ z_j\geq 0\quad (j=1,2,3,4) \\ \underline{\omega}^{\mathrm{LNZ}} \text{ and } \bar{\omega}^{\mathrm{LNZ}} \text{ unrestricted in sign,} \end{cases} \tag{1.71}$$

where z_j $(j=1,2,3,4)$, $\underline{\omega}^{\mathrm{LNZ}}$, and $\bar{\omega}^{\mathrm{LNZ}}$ are decision variables.

Still taking $\varepsilon=0.5$, and solving Eq. (1.71) by using the simplex method of linear programming, we can obtain its optimal solution whose components are given as follows:

$$\underline{\omega}^{\mathrm{LNZ}*}=0.054, \bar{\omega}^{\mathrm{LNZ}*}=0.141, z_1^*=0, z_2^*=0, z_3^*=0.852, z_4^*=0.148.$$

Hence, we obtain the optimal mixed strategy $z^*=(z_1^*, z_2^*, z_3^*, z_4^*)^{\mathrm{T}}=(0,0,0.852,0.148)^{\mathrm{T}}$ and the loss-ceiling $[\underline{\mu}^{\mathrm{LNZ}}, \bar{\mu}^{\mathrm{LNZ}}]=[\underline{\omega}^{\mathrm{LNZ}*}, \bar{\omega}^{\mathrm{LNZ}*}]=[0.054, 0.141]$ for the player II.

Obviously, we have the following interval inclusion relation:

$$[\underline{\nu}^{\mathrm{LNZ}}, \bar{\nu}^{\mathrm{LNZ}}]=[0.068, 0.127]\subset[0.054, 0.141]=[\underline{\mu}^{\mathrm{LNZ}}, \bar{\mu}^{\mathrm{LNZ}}].$$

Furthermore, according to Definition 1.3, we can easily obtain the acceptability index of the above two intervals' comparison as follows:

$$\begin{aligned}\varphi([\underline{\nu}^{\mathrm{LNZ}}, \bar{\nu}^{\mathrm{LNZ}}]\leq_I[\underline{\mu}^{\mathrm{LNZ}}, \bar{\mu}^{\mathrm{LNZ}}]) &= \frac{\bar{\mu}^{\mathrm{LNZ}}-\bar{\nu}^{\mathrm{LNZ}}}{(\bar{\mu}^{\mathrm{LNZ}}-\underline{\mu}^{\mathrm{LNZ}})-(\bar{\nu}^{\mathrm{LNZ}}-\underline{\nu}^{\mathrm{LNZ}})} \\ &= \frac{0.141-0.127}{(0.141-0.054)-(0.127-0.068)}=0.5,\end{aligned}$$

i.e., the interval-valued inequality (or order relation) $[\underline{\nu}^{\mathrm{LNZ}}, \bar{\nu}^{\mathrm{LNZ}}]\leq_I[\underline{\mu}^{\mathrm{LNZ}}, \bar{\mu}^{\mathrm{LNZ}}]$ (i.e., $[0.068, 0.127]\leq_I[0.054, 0.141]$) is valid with the acceptability degree 0.5. In other words, the statement "the player I's gain-floor is not larger than the player II's loss-ceiling" is true with the acceptability degree 0.5.

Analogously, for other specific given values of the parameter $\varepsilon\in[0,1]$, we can solve Eqs. (1.70) and (1.71) and hereby obtain the player I's gain-floor and the player II's loss-ceiling as well as their corresponding optimal strategies (omitted).

(2e) Computational results obtained by Shashikhin's method

According to the interval arithmetic [20], Shashikhin [19] defined the generalized minimum and maximum operators of the intervals and hereby suggested $\max_{1\leq i\leq m}\min_{1\leq j\leq n}\{[a_{Lij}, a_{Rij}]\}$ and $\min_{1\leq j\leq n}\max_{1\leq i\leq m}\{[a_{Lij}, a_{Rij}]\}$ as the player I's gain-floor and the player II's loss-ceiling in the interval-valued matrix game $\bar{\boldsymbol{A}} = ([a_{Lij}, a_{Rij}])_{m\times n}$, respectively.

For the above interval-valued matrix game $\bar{\boldsymbol{A}}_0$, according to Shashikhin's method [19], we can readily obtain

$$\begin{aligned}\min_{1\leq j\leq 4}\{[a_{L1j}, a_{R1j}]\} &= \min\{[0.12, 0.17], [0.11, 0.16], [0.075, 0.12], \\ &\qquad [0.068, 0.13]\} = [0.068, 0.12], \\ \min_{1\leq j\leq 4}\{[a_{L2j}, a_{R2j}]\} &= \min\{[0.18, 0.22], [0.12, 0.15], [0.072, 0.14], \\ &\qquad [-0.05, 0.15]\} = [-0.05, 0.14]\end{aligned}$$

and

$$\begin{aligned}\min_{1\leq j\leq 4}\{[a_{L3j}, a_{R3j}]\} &= \min\{[0.043, 0.043], [0.043, 0.043], [0.043, 0.043], \\ &\qquad [0.043, 0.043]\} = [0.043, 0.043].\end{aligned}$$

Hence, we have

$$\begin{aligned}\max_{1\leq i\leq 3}\min_{1\leq j\leq 4}\{[a_{Lij}, a_{Rij}]\} &= \max\{[0.068, 0.12], [-0.05, 0.14], [0.043, 0.043]\} \\ &= [0.068, 0.14],\end{aligned}$$

i.e., the value of the interval-valued matrix game $\bar{\boldsymbol{A}}_0$ for the investor (i.e., player I) is a closed interval $[\underline{v}^{\mathrm{S}}, \bar{v}^{\mathrm{S}}] = [0.068, 0.140]$.

Similarly, we can easily obtain

$$\begin{aligned}\max_{1\leq i\leq 3}\{[a_{Li1}, a_{Ri1}]\} &= \max\{[0.12, 0.17], [0.18, 0.22], [0.043, 0.043]\} = [0.18, 0.22], \\ \max_{1\leq i\leq 3}\{[a_{Li2}, a_{Ri2}]\} &= \max\{[0.11, 0.16], [0.12, 0.15], [0.043, 0.043]\} = [0.12, 0.16], \\ \max_{1\leq i\leq 3}\{[a_{Li3}, a_{Ri3}]\} &= \max\{[0.075, 0.12], [0.072, 0.14], [0.043, 0.043]\} = [0.075, 0.14]\end{aligned}$$

and

$$\max_{1 \le i \le 3}\{[a_{Li4}, a_{Ri4}]\} = \max\{[0.068, 0.13], [-0.05, 0.15], [0.043, 0.043]\}$$
$$= [0.068, 0.15].$$

Hence, we have

$$\min_{1 \le j \le 4} \max_{1 \le i \le 3}\{[a_{Lij}, a_{Rij}]\} = \min\{[0.18, 0.22], [0.12, 0.16], [0.075, 0.14], [0.068, 0.15]\}$$
$$= [0.068, 0.14],$$

i.e., the value of the interval-valued matrix game $\bar{\boldsymbol{A}}_0$ for the economic environment/nature (i.e., player II) is a closed interval $[\underline{\mu}^S, \bar{\mu}^S] = [0.068, 0.14]$.

Therefore, we can have

$$\max_{1 \le i \le 3} \min_{1 \le j \le 4}\{[a_{Lij}, a_{Rij}]\} = \min_{1 \le j \le 4} \max_{1 \le i \le 3}\{[a_{Lij}, a_{Rij}]\} = [0.068, 0.14].$$

Then, the value of the interval-valued matrix game $\bar{\boldsymbol{A}}_0$ is a closed interval $[\underline{V}^S, \bar{V}^S] = [0.068, 0.14]$.

(3) The obtained results' comparison and conclusions

It is not difficult to draw the following conclusions from the aforementioned modeling, solving process and computational results.

(3a) The linear programming method proposed in this section, Liu and Kao's method [18], and Shashikhin's method [19] obtain the identical interval-type value $[0.068, 0.14]$ of the interval-valued matrix game $\bar{\boldsymbol{A}}_0$. Li's model [29] obtains only the values $[\underline{v}^L, \bar{v}^L] = [0.068, 0.12]$ and $[\underline{\mu}^L, \bar{\mu}^L] = [0.075, 0.14]$ of the interval-valued matrix game $\bar{\boldsymbol{A}}_0$ for the players I and II, respectively. The weighted average method [28] obtains only the values $[\underline{v}^{LNZ}, \bar{v}^{LNZ}] = [0.068, 0.127]$ and $[\underline{\mu}^{LNZ}, \bar{\mu}^{LNZ}] = [0.054, 0.141]$ of the interval-valued matrix game $\bar{\boldsymbol{A}}_0$ for the players I and II with the given acceptability degree 0.5 *a priori*, respectively. Moreover, the interval-valued inequalities (or order relations) $[\underline{v}^L, \bar{v}^L] \le_I [\underline{\mu}^L, \bar{\mu}^L]$ and $[\underline{v}^{LNZ}, \bar{v}^{LNZ}] \le_I [\underline{\mu}^{LNZ}, \bar{\mu}^{LNZ}]$ are valid with the acceptability degrees 1 and 0.5, respectively. From the concepts of the "zero-sum" and the value of the interval-valued matrix game, however, the computational results obtained by the linear programming method proposed in this section, Liu and Kao's method [18], and Shashikhin's method [19] are more rational, reliable, and convinced than those obtained by Li's model [29] and the weighted average method [28].

(3b) The linear programming method in this section is developed on the monotonicity of the value of the interval-valued matrix game and the duality theorem of linear programming. Liu and Kao's method [18] used the duality theorem of linear programming and a variable substitution technique to construct the auxiliary linear programming models. These two methods always ensure that any interval-valued matrix game has a value. Moreover, they are not involved in any subjective factor. Li's model [29] employed the interval comparison relation to establish the two-level linear programming method. Shashikhin's method [19] used the generalized minimum and maximum operators of the intervals to define the player I's gain-floor and the player II's loss-ceiling. The weighted average method [28] as well as the similar methods [24–26, 31] used the acceptability index of the interval comparison operator (or inequalities) to construct crisply equivalent mathematical programming models. These methods [24–26, 28, 31] closely depend on interval comparison operators (or order relations), which are difficult to be appropriately determined. In addition, they cannot always ensure that any interval-valued matrix game has a value. In other words, they usually obtain only the values of the interval-valued matrix game for the players I and II.

(3c) The amount of computation and complexity of solving process for the linear programming method in this section are less than those of Kao's method [18], Li's model [29], and the weighted average method [28] as well as the similar methods [24–26, 31] since these latter methods usually result in more additional variables, constraints (equalities and inequalities), and unrestricted/restricted in sign in the constructed mathematical programming models and more auxiliary mathematical programming models, which need to be solved. In addition, the additional constraints may be superabundant and even contradictable. Solving process of Shashikhin's method [19] is simple. Stated as earlier, however, Shashikhin's method [19] cannot always ensure that any interval-valued matrix game has a value.

References

1. Owen G (1982) Game theory, 2nd edn. Academic Press, New York
2. Nishizaki I, Sakawa M (2001) Fuzzy and multiobjective games for conflict resolution. Physica-Verlag, Springer-Verlag, Berlin
3. Li D-F (2003) Fuzzy multiobjective many-person decision makings and games. National Defense Industry Press, Beijing (in Chinese)
4. Bector CR, Chandra S (2005) Fuzzy mathematical programming and fuzzy matrix games. Springer-Verlag, Berlin
5. Dubois D, Prade H (1980) Fuzzy sets and systems: theory and applications. Academic Press, New York
6. Bector CR, Chandra S, Vijay V (2004) Matrix games with fuzzy goals and fuzzy linear programming duality. Fuzzy Optim Decis Making 3:255–269
7. Bector CR, Chandra S, Vijay V (2004) Duality in linear programming with fuzzy parameters and matrix games with fuzzy pay-offs. Fuzzy Sets Syst 46(2):253–269

8. Campos L (1989) Fuzzy linear programming models to solve fuzzy matrix games. Fuzzy Sets Syst 32:275–289
9. Campos L, Gonzalez A (1991) Fuzzy matrix games considering the criteria of the players. Kybernetes 20:17–23
10. Campos L, Gonzalez A, Vila MA (1992) On the use of the ranking function approach to solve fuzzy matrix games in a direct way. Fuzzy Sets Syst 49:193–203
11. Maeda T (2003) On characterization of equilibrium strategy of two-person zero-sum games with fuzzy payoffs. Fuzzy Sets Syst 139:283–296
12. Li D-F (2011) Linear programming approach to solve interval-valued matrix games. Omega: Int J Manage Sci 39(6):655–666
13. Nishizaki I, Sakawa M (2000) Equilibrium solutions in multiobjective bi-matrix games with fuzzy payoffs and fuzzy goals. Fuzzy Sets Syst 111(1):99–116
14. Nishizaki I, Sakawa M (2000) Solutions based on fuzzy goals in fuzzy linear programming games. Fuzzy Sets Syst 115(1):105–119
15. Sakawa M, Nishizaki I (1994) Max-min solutions for fuzzy multiobjective matrix games. Fuzzy Sets Syst 67:53–69
16. Vijay V, Chandra S, Bector CR (2005) Matrix games with fuzzy goals and fuzzy payoffs. Omega: Int J Manage Sci 33:425–429
17. Larbani M (2009) Non cooperative fuzzy games in normal form: a survey. Fuzzy Sets Syst 160:3184–3210
18. Liu S-T, Kao C (2009) Matrix games with interval data. Comput Ind Eng 56(4):1697–1700
19. Shashikhin VN (2004) Antagonistic game with interval payoff functions. Cybern Syst Anal 40 (4):556–564
20. Moore RE (1979) Method and application of interval analysis. SIAM, Philadelphia
21. Ramadan K (1996) Linear programming with interval coefficients, M. Sc. Thesis, Carleton University, Ottawa, Ontario, Canada
22. Sengupta A, Pal TK (2000) Theory and methodology on comparing interval numbers. Eur J Oper Res 127:28–43
23. Sengupta A, Pal TK, Chakraborty D (2001) Interpretation of inequality constraints involving interval coefficients and a solution to interval programming. Fuzzy Sets Syst 119:129–138
24. Collins WD, Hu C-Y (2008) Studying interval valued matrix games with fuzzy logic. Soft Comput 12(2):147–155
25. Nayak PK, Pal M (2009) Linear programming technique to solve two person matrix games with interval pay-offs. Asia-Pac J Oper Res 26(2):285–305
26. Li D-F (2011) Notes on "Linear programming technique to solve two person matrix games with interval pay-offs". Asia-Pac J Oper Res 28(6):705–737
27. Li D-F (2008) Lexicographic method for matrix games with payoffs of triangular fuzzy numbers. Int J Uncertainty Fuzziness and Knowl Based Syst 16(3):371–389
28. Li D-F, Nan J-X, Zhang M-J (2012) Interval programming models for matrix games with interval payoffs. Optim Methods Softw 27(1):1–16
29. Li D-F (1999) A fuzzy multiobjective programming approach to solve fuzzy matrix games. J Fuzzy Math 7(4):907–912
30. Zadeh L (1965) Fuzzy sets. Inf Control 8:338–356
31. Collins WD, Hu C-Y (2008) Interval matrix games. In: Hu C-Y, Kearfott RB, Korvinet AD et al (eds) Knowledge processing with interval and soft computing. Springer, London, pp 168–172
32. Ishibuchi H, Tanaka H (1990) Multiobjective programming in optimization of the interval objective function. Eur J Oper Res 48:219–225

33. Chanas S, Kuchta D (1996) Multiobjective programming in optimization of interval objective functions—a generalized approach. Eur J Oper Res 94:594–598
34. Tong S (1994) Interval number and fuzzy number linear programming. Fuzzy Sets Syst 66:301–306
35. Li D-F, Yang J-B (2004) Two level linear programming approach to solve fuzzy matrix games with fuzzy payoffs. University of Manchester Institute of Science and Technology, UK, Unpublished preprint, Manchester School of Management

Chapter 2
Matrix Games with Payoffs of Triangular Fuzzy Numbers

2.1 Introduction

The matrix game theory gives a mathematical background for dealing with competitive or antagonistic situations arise in many parts of real life. Matrix games have been extensively studied and successfully applied to many fields such as economics, business, management, and e-commerce as well as advertising. As stated in Chap. 1, however, the assumption that all payoffs are precise common knowledge to both the players is not realistic in many antagonistic decision occasions. In fact, more often than not, in real antagonistic situations, the players are not able to exactly estimate payoffs in the game due to lack of adequate information and/or imprecision of the available information on the environment [1, 2]. This lack of precision and certainty may be appropriately modeled by using the fuzzy set [3–6]. As a special case of fuzzy sets, intervals which are also called fuzzy intervals or interval-valued fuzzy sets are used to deal with fuzziness in matrix games. Consequently, we have extensively studied interval-valued matrix games. From now on, we focus on studying fuzzy matrix games with payoffs represented by fuzzy numbers such as triangular fuzzy numbers and trapezoidal fuzzy numbers.

Fuzzy matrix games were firstly solved by developing the fuzzy linear programming method based on ranking functions of fuzzy numbers and auxiliary linear programming models [7–9]. However, Campos' methods [7–9] provided only crisp solutions with interpretation of fuzzy semantics. Their results were generalized to multi-objective matrix games with fuzzy payoffs and fuzzy goals [10, 11]. Bector and Chandra [12], Bector et al. [13, 14], and Vijay et al. [15] proposed linear programming methods for solving fuzzy matrix games based on certain duality for linear programming with fuzzy parameters. These works cannot provide membership functions of the gain-floor and loss-ceiling for the players even though they are very much desirable. The above methods were essentially the same as that of

D.-F. Li, *Linear Programming Models and Methods of Matrix Games with Payoffs of Triangular Fuzzy Numbers*, Studies in Fuzziness and Soft Computing 328 ,
DOI 10.1007/978-3-662-48476-0_2

Campos [7] but certain modifications were made to help in having a better understanding of the same. Obviously, all the aforementioned methods are defuzzification ones based on suitable ranking functions, which are not easily chosen. In these methods, the obtained solutions closely depend on ranking functions and more or less involve in subjective factors such as attitudes and preference. On the other hand, these methods provided only defuzzification ones of the gain-floor and loss-ceiling for the players, whose membership functions cannot be explicitly obtained even though they are very much desirable. Moreover, it is not always sure that the obtained defuzzification gain-floor and loss-ceiling for the players are identical. This case is not rational and effective. From viewpoints of logic and the concept of matrix games with fuzzy payoffs, the gain-floor and loss-ceiling for the players should be fuzzy and identical since the expected payoffs are a linear combination of fuzzy payoffs and the matrix games are zero-sum.

Li [16] (with reference to [17]) proposed the two-level linear programming method for solving matrix games with payoffs of triangular fuzzy numbers, which was called as Li's model by Bector and Chandra [12] and Larbani [18]. In Li's model [16], the obtained gain-floor and loss-ceiling for the players are fuzzy and their membership functions can be explicitly obtained. However, Li's model cannot always guarantee that the gain-floor and loss-ceiling for the players are identical and hereby any fuzzy matrix game with payoffs of triangular fuzzy numbers has a fuzzy value, which is not rational since the matrix game is zero-sum. As far as we know, there is no method which can always guarantee that the gain-floor and loss-ceiling for the players are identical and hereby the matrix game with fuzzy payoffs has a fuzzy value, whose membership functions can be explicitly obtained. In this chapter, we will focus on studying matrix games with payoffs of triangular fuzzy numbers. Selecting triangular fuzzy numbers to express fuzzy payoffs stems from the fact that in many management applications they provide a very convenient object for the representation of imprecision and uncertain information in payoffs. On the one hand, triangular fuzzy numbers allow the modeling of a wide class of fuzzy numbers. Intervals and real numbers are special cases of triangular fuzzy numbers. On the other hand, triangular fuzzy numbers are easily extended to trapezoidal fuzzy numbers. Using triangular fuzzy numbers, we also have the freedom of being or not being symmetric. Another positive feature of the triangular fuzzy numbers is the ease of acquiring the necessary parameters. An additional consideration in using the triangular fuzzy number is the ease with which it can be manipulated in the context of the application.

In this chapter, we will propose some important concepts of solutions of matrix games with payoffs of triangular fuzzy numbers and develop auxiliary linear programming models and methods for solving matrix games with payoffs of triangular fuzzy numbers. Stated as earlier, it is easy to see that some linear programming models and methods proposed in this chapter are easily extended to establish those for matrix games with payoffs of trapezoidal fuzzy numbers.

2.2 Triangular Fuzzy Numbers and Alfa-Cut Sets

A fuzzy number $\tilde{a}$ with the membership function $\mu_{\tilde{a}}(x)$ is a special fuzzy subset of the real number set R, which satisfies the following two conditions [3]:

1. there exists at least a real number $x_0 \in \mathrm{R}$ so that $\mu_{\tilde{a}}(x_0) = 1$;
2. the membership function $\mu_{\tilde{a}}(x)$ is left and right continuous, depicted as in Fig. 2.1.

In the following, we mainly review a special and an important forms of fuzzy numbers: triangular fuzzy numbers.

Triangular fuzzy numbers are a special case of fuzzy numbers. A triangular fuzzy number $\tilde{a} = (a^l, a^m, a^r)$ is a special fuzzy number [3], whose membership function is given as follows:

$$\mu_{\tilde{a}}(x) = \begin{cases} \frac{x-a^l}{a^m-a^l} & \text{if } a^l \leq x < a^m \\ 1 & \text{if } x = a^m \\ \frac{a^r-x}{a^r-a^m} & \text{if } a^m < x \leq a^r \\ 0 & \text{else,} \end{cases} \tag{2.1}$$

where a^m is the mean of $\tilde{a}$, a^l and a^r are the lower and upper limits (bounds) of $\tilde{a}$, respectively, depicted as in Fig. 2.2. The set of triangular fuzzy numbers is denoted by T(R).

Obviously, if $a^l = a^m = a^r$, then the triangular fuzzy number $\tilde{a} = (a^l, a^m, a^r)$ is reduced to a real number. Conversely, a real number is easily rewritten as a

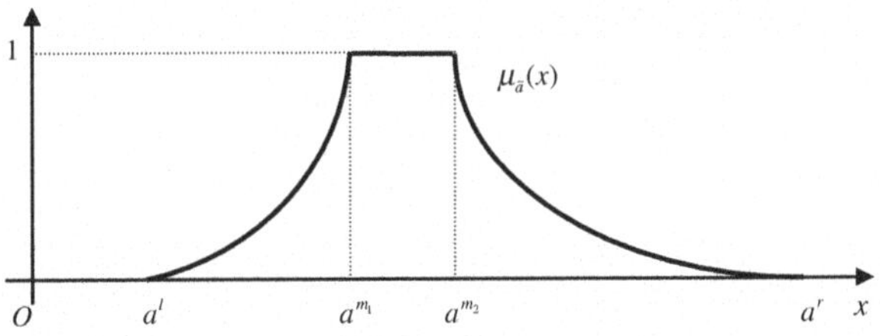

Fig. 2.1 A fuzzy number

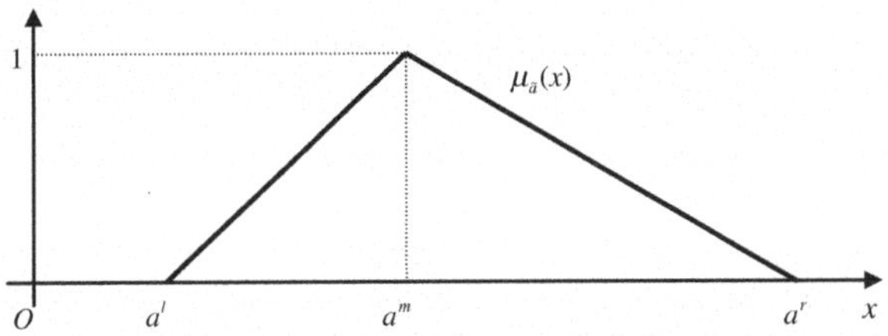

Fig. 2.2 A triangular fuzzy number

triangular fuzzy number. Thus, the triangular fuzzy number can be flexible to represent various semantics of uncertainty such as ill-quantity [5].

If $a^l \geq 0$ and $a^r > 0$, then $\tilde{a} = (a^l, a^m, a^r)$ is called a non-negative triangular fuzzy number, denoted by $\tilde{a} \geq 0$. If $a^l > 0$, then $\tilde{a}$ is called a positive triangular fuzzy number, denoted by $\tilde{a} > 0$. Conversely, if $a^r \leq 0$ and $a^l < 0$, then $\tilde{a}$ is called a non-positive triangular fuzzy number, denoted by $\tilde{a} \leq 0$. If $a^r < 0$, then $\tilde{a}$ is called a negative triangular fuzzy number, denoted by $\tilde{a} < 0$.

Let $\tilde{a} = (a^l, a^m, a^r)$ and $\tilde{b} = (b^l, b^m, b^r)$ be two triangular fuzzy numbers. Then, their arithmetical operations can be expressed as follows:

$$\tilde{a} + \tilde{b} = (a^l + b^l, a^m + b^m, a^r + b^r) \tag{2.2}$$

and

$$\lambda\tilde{a} = \begin{cases} (\lambda a^l, \lambda a^m, \lambda a^r) & \text{if } \lambda \geq 0 \\ (\lambda a^r, \lambda a^m, \lambda a^l) & \text{if } \lambda < 0, \end{cases} \tag{2.3}$$

where $\lambda \in \mathrm{R}$ is a real number.

A α-cut set of the triangular fuzzy number $\tilde{a} = (a^l, a^m, a^r)$ is defined as $\tilde{a}(\alpha) = \{x | \mu_{\tilde{a}}(x) \geq \alpha\}$, where $\alpha \in [0, 1]$. Thus, for any $\alpha \in [0, 1]$, we can obtain a α-cut set of the triangular fuzzy number $\tilde{a}$, which is an interval, denoted by $\tilde{a}(\alpha) = [a^L(\alpha), a^R(\alpha)]$. It is easily derived from Eq. (2.1) that

$$a^L(\alpha) = \alpha a^m + (1 - \alpha)a^l$$

and

$$a^R(\alpha) = \alpha a^m + (1 - \alpha)a^r.$$

In particular, we have

$$\tilde{a}(1) = [a^L(1), a^R(1)] = [a^m, a^m] = a^m$$

and

$$\tilde{a}(0) = [a^L(0), a^R(0)] = [a^l, a^r].$$

According to the operations over intervals [19], we can easily have:

$$[a^L(\alpha), a^R(\alpha)] = \alpha[a^m, a^m] + (1 - \alpha)[a^l, a^r] = \alpha\tilde{a}(1) + (1 - \alpha)\tilde{a}(0), \tag{2.4}$$

which means that any α-cut set of an arbitrary triangular fuzzy number can be directly obtained from its 1-cut set and 0-cut set, depicted as in Fig. 2.3.

According to the representation theorem for the fuzzy set [5], using Eq. (2.4), any triangular fuzzy number $\tilde{a} = (a^l, a^m, a^r)$ can be expressed as follows:

Fig. 2.3 α-cut sets of a triangular fuzzy number

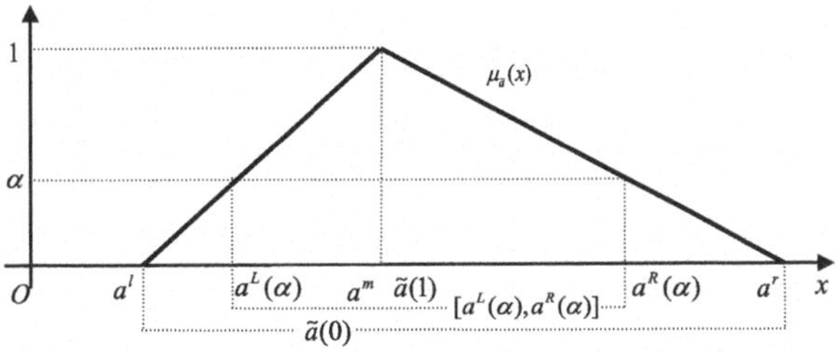

$$\tilde{a} = \bigcup_{\alpha \in [0,1]} \{\alpha \otimes \tilde{a}(\alpha)\} = \bigcup_{\alpha \in [0,1]} \{\alpha \otimes [\alpha \tilde{a}(1) + (1-\alpha)\tilde{a}(0)]\}, \quad (2.5)$$

where $\alpha \otimes \tilde{a}(\alpha)$ is defined as a fuzzy set, whose membership function is given as follows:

$$\mu_{\alpha \otimes \tilde{a}(\alpha)}(x) = \begin{cases} \alpha & \text{if } x \in \tilde{a}(\alpha) \\ 0 & \text{otherwise.} \end{cases}$$

Equation (2.5) means that any triangular fuzzy number can be directly constructed through using its 1-cut set and 0-cut set.

From the aforementioned discussion, we summarize the conclusion as in Theorem 2.1, which will be used to construct the fuzzy values of matrix games with payoffs of triangular fuzzy numbers.

Theorem 2.1 *Any triangular fuzzy number and its α-cuts have the relations (1) and (2) as follows:*

1. *Any α-cut of a triangular fuzzy number can be directly obtained from both its 1-cut and 0-cut;*
2. *Any triangular fuzzy number can be directly constructed by using both its 1-cut and 0-cut.*

Proof According to the concept of α-cuts of triangular fuzzy numbers and the representation theorem for the fuzzy set, it is easy to prove that (1) and (2) of Theorem 2.1 are valid (omitted).

2.3 Fuzzy Multi-Objective Programming Models of Matrix Games with Payoffs of Triangular Fuzzy Numbers

2.3.1 Order Relations of Triangular Fuzzy Numbers

In contrast with the intervals' ranking or order relation as stated in Sects. 1.3 and 1.4, it is very difficult to rank (or compare) fuzzy numbers. Ramik and Rimanek [20] gave the definition of the order relation "$\tilde{\le}$" for general fuzzy numbers. In this section, the order relations "$\tilde{\le}$" and "$\tilde{\ge}$" are used only for triangular fuzzy numbers, not for general fuzzy numbers as stated in Sect. 2.2. To be more precisely, we give the meaning of the order relations "$\tilde{\le}$" and "$\tilde{\ge}$" for the triangular fuzzy numbers in Definition 2.1 as follows.

Definition 2.1 Let $\tilde{a} = (a^l, a^m, a^r)$ and $\tilde{b} = (b^l, b^m, b^r)$ be two triangular fuzzy numbers. Then, $\tilde{a} \tilde{\le} \tilde{b}$ if and only if $a^l \le b^l$, $a \le b$, and $a^r \le b^r$. Similarly, $\tilde{a} \tilde{\ge} \tilde{b}$ if and only if $a^l \ge b^l$, $a \ge b$, and $a^r \ge b^r$.

The validity of Definition 2.1 may be discussed in a similar way to that of fuzzy numbers [20].

"$\tilde{\le}$" and "$\tilde{\ge}$" are fuzzy versions of the order relations "$\le$" and "$\ge$" in the three-dimension Euclidean space R^3, and have the linguistic interpretation "essentially less than or equal to" and "essentially greater than or equal to", respectively.

Analogously, $\tilde{a} \tilde{<} \tilde{b}$ if and only if $\tilde{a} \tilde{\le} \tilde{b}$ and $\tilde{a} \ne \tilde{b}$. $\tilde{a} \tilde{>} \tilde{b}$ if and only if $\tilde{a} \tilde{\ge} \tilde{b}$ and $\tilde{a} \ne \tilde{b}$.

From Definition 2.1, a triangular fuzzy number $\tilde{a} \in T(\mathrm{R})$ may be regarded as a three-dimension vector and the order relations "$\tilde{\le}$" and "$\tilde{\ge}$" are similar to those in the three-dimension Euclidean space R^3. Thus, the definition of maximizing and minimizing triangular fuzzy numbers can be given as follows.

Definition 2.2 Let $\tilde{a} = (a^l, a^m, a^r)$ be any triangular fuzzy number. A maximization problem of triangular fuzzy numbers is expressed as follows:

$$\max\{\tilde{a} | \tilde{a} \in \Omega_3 \cap T(\mathrm{R})\},$$

which is equivalent to the multi-objective mathematical programming model as follows:

$$\begin{aligned} &\max\{a^l\} \\ &\max\{a^m\} \\ &\max\{a^r\} \\ &\text{s.t.} \begin{cases} \tilde{a} \in \Omega_3 \\ a^l \le a^m \le a^r \\ a^l, a^m, \text{ and } a^r \text{ unrestricted in sign,} \end{cases} \end{aligned}$$

where $T(\mathrm{R})$ is the set of triangular fuzzy numbers as stated in Sect. 2.2, Ω_3 is the set of constraints in which the variable $\tilde{a}$ should be satisfied according to requirements in the real situation.

Definition 2.3 Let $\tilde{a} = (a^l, a^m, a^r)$ be any triangular fuzzy number. A minimization problem of triangular fuzzy numbers is described as follows:

$$\min\{\tilde{a}|\tilde{a} \in \Omega_4 \cap T(\mathrm{R})\},$$

which is equivalent to the multi-objective mathematical programming model as follows:

$$\begin{aligned}
&\min\{a^l\}\\
&\min\{a^m\}\\
&\min\{a^r\}\\
&\text{s.t.} \begin{cases} \tilde{a} \in \Omega_4 \\ a^l \leq a^m \leq a^r \\ a^l, a^m, \text{ and } a^r \text{ unrestricted in sign,} \end{cases}
\end{aligned}$$

where Ω_4 is the set of constraints in which the variable $\tilde{a}$ should be satisfied according to requirements in the real situation.

Definitions 2.2 and 2.3 can be used to transform corresponding fuzzy optimization problems of matrix games with payoffs of triangular fuzzy numbers into multi-objective linear programming models, which may be solved by using the existing multi-objective programming methods [21, 22].

2.3.2 *Concepts of Solutions of Matrix Games with Payoffs of Triangular Fuzzy Numbers*

Let us consider matrix games with payoffs of triangular fuzzy numbers, where the sets of pure strategies and the sets of mixed strategies for the players I and II respectively are S_1, S_2, Y, and Z defined as in Sect. 1.2. Assume that the payoff matrix of the player I is given as follows:

$$\tilde{\boldsymbol{A}} = (\tilde{a}_{ij})_{m\times n} = \begin{matrix} \\ \delta_1 \\ \delta_2 \\ \vdots \\ \delta_m \end{matrix} \begin{matrix} \begin{matrix} \beta_1 & \beta_2 & \cdots & \beta_n \end{matrix} \\ \begin{pmatrix} \tilde{a}_{11} & \tilde{a}_{12} & \cdots & \tilde{a}_{1n} \\ \tilde{a}_{21} & \tilde{a}_{22} & \cdots & \tilde{a}_{2n} \\ \vdots & \vdots & \cdots & \vdots \\ \tilde{a}_{m1} & \tilde{a}_{m2} & \cdots & \tilde{a}_{mn} \end{pmatrix} \end{matrix},$$

where $\tilde{a}_{ij} = (a^l_{ij}, a^m_{ij}, a^r_{ij})$ $(i = 1, 2, \ldots, m;\ j = 1, 2, \ldots, n)$ are triangular fuzzy numbers defined as in Sect. 2.2. Then, a matrix game with payoffs of triangular fuzzy numbers is expressed with $\tilde{\boldsymbol{A}}$ for short.

According to Eqs. (2.2) and (2.3), the fuzzy expected payoff (or value) of the player I can be computed as follows:

$$\tilde{E}(\tilde{\boldsymbol{A}}) = \boldsymbol{y}^{\mathrm{T}}\tilde{\boldsymbol{A}}\boldsymbol{z} = \sum_{i=1}^{m}\sum_{j=1}^{n}\tilde{a}_{ij}y_i z_j = \left(\sum_{i=1}^{m}\sum_{j=1}^{n}a_{ij}^{l}y_i z_j, \sum_{i=1}^{m}\sum_{j=1}^{n}a_{ij}^{m}y_i z_j, \sum_{i=1}^{m}\sum_{j=1}^{n}a_{ij}^{r}y_i z_j\right),$$

which is a triangular fuzzy number.

As the matrix game $\tilde{\boldsymbol{A}}$ with payoffs of triangular fuzzy numbers is zero-sum, according to Eq. (2.3), the fuzzy expected payoff of the player II is equal to

$$\begin{aligned}\tilde{E}(-\tilde{\boldsymbol{A}}) &= \boldsymbol{y}^{\mathrm{T}}(-\tilde{\boldsymbol{A}})\boldsymbol{z} = \sum_{i=1}^{m}\sum_{j=1}^{n}(-\tilde{a}_{ij})y_i z_j \\ &= \left(-\sum_{i=1}^{m}\sum_{j=1}^{n}a_{ij}^{r}y_i z_j, -\sum_{i=1}^{m}\sum_{j=1}^{n}a_{ij}^{m}y_i z_j, -\sum_{i=1}^{m}\sum_{j=1}^{n}a_{ij}^{l}y_i z_j\right),\end{aligned}$$

which is also a triangular fuzzy number. Thus, in general, the player I's gain-floor and the player II's loss-ceiling should be triangular fuzzy numbers, denoted by $\tilde{\upsilon} = (\upsilon^l, \upsilon^m, \upsilon^r)$ and $\tilde{\omega} = (\omega^l, \omega^m, \omega^r)$, respectively.

Since the fuzzy expected payoffs of the players and the player I's gain-floor and the player II's loss-ceiling are triangular fuzzy numbers, thus according to Definitions 2.2 and 2.3, the concept of solutions of matrix games with payoffs of triangular fuzzy numbers may be given by using that of the Pareto optimal solution as follows. Bector et al. [13, 14] firstly introduced the notion of reasonable solutions of fuzzy matrix games, which is a generalization of that of fuzzy matrix games [23].

Definition 2.4 Let $\tilde{\upsilon} = (\upsilon^l, \upsilon^m, \upsilon^r)$ and $\tilde{\omega} = (\omega^l, \omega^m, \omega^r)$ be triangular fuzzy numbers. Assume that there exist mixed strategies $\boldsymbol{y}^* \in Y$ and $\boldsymbol{z}^* \in Z$. Then, $(\boldsymbol{y}^*, \boldsymbol{z}^*, \tilde{\upsilon}, \tilde{\omega})$ is called a reasonable solution of the matrix game $\tilde{\boldsymbol{A}}$ with payoffs of triangular fuzzy numbers if it satisfies both the following conditions:

1. $\boldsymbol{y}^{*\mathrm{T}}\tilde{\boldsymbol{A}}\boldsymbol{z} \tilde{\geq} \tilde{\upsilon}$
and
2. $\boldsymbol{y}^{\mathrm{T}}\tilde{\boldsymbol{A}}\boldsymbol{z}^* \tilde{\leq} \tilde{\omega}$
for any $\boldsymbol{z} \in Z$ and $\boldsymbol{y} \in Y$.

If $(\boldsymbol{y}^*, \boldsymbol{z}^*, \tilde{\upsilon}, \tilde{\omega})$ is a reasonable solution of the matrix game $\tilde{\boldsymbol{A}}$ with payoffs of triangular fuzzy numbers, then $\tilde{\upsilon}$ and $\tilde{\omega}$ are called reasonable values for the players I and II, $\boldsymbol{y}^*$ and $\boldsymbol{z}^*$ are called reasonable (mixed) strategies for the players I and II, respectively.

The sets of all reasonable values $\tilde{\upsilon}$ and $\tilde{\omega}$ for the players I and II are denoted by U and W, respectively.

As stated earlier, Definition 2.4 only gives the notion of reasonable solutions of matrix games with payoffs of triangular fuzzy numbers rather than the notion of optimal solutions. Thus, we give the concept of solutions of matrix games with payoffs of triangular fuzzy numbers as in the following Definition 2.5.

Definition 2.5 Assume that there exist $\tilde{\upsilon}^* \in U$ and $\tilde{\omega} \in W$. If there do not exist any $\tilde{\upsilon} \in U$ $(\tilde{\upsilon} \neq \tilde{\upsilon}^*)$ and $\tilde{\omega} \in W$ $(\tilde{\omega} \neq \tilde{\omega}^*)$ so that

1. $\tilde{\upsilon}^* \tilde{\leq} \tilde{\upsilon}$
and
2. $\tilde{\omega}^* \tilde{\geq} \tilde{\omega}$,

then, $(\boldsymbol{y}^*, \boldsymbol{z}^*, \tilde{\upsilon}^*, \tilde{\omega}^*)$ is called a solution of the matrix game $\tilde{\boldsymbol{A}}$ with payoffs of triangular fuzzy numbers, $\boldsymbol{y}^*$ and $\boldsymbol{z}^*$ are called a maximin (mixed) strategy and a minimax (mixed) strategy for the players I and II, $\tilde{\upsilon}^*$ and $\tilde{\omega}^*$ are called the player I's gain-floor and the player II's loss-ceiling (or fuzzy values for the players I and II), respectively.

Let

$$\tilde{V}^* = \tilde{\upsilon}^* \wedge \tilde{\omega}^*$$

with the membership function

$$\mu_{\tilde{V}^*}(x) = \min_x \{\mu_{\tilde{\upsilon}^*}(x), \mu_{\tilde{\omega}^*}(x)\}.$$

Then, $\tilde{V}^*$ is called a fuzzy equilibrium value of the matrix game $\tilde{\boldsymbol{A}}$ with payoffs of triangular fuzzy numbers, depicted as in Fig. 2.4.

It is easy to see from Fig. 2.4 that a fuzzy value $\tilde{V}^*$ of the matrix game $\tilde{\boldsymbol{A}}$ with payoffs of triangular fuzzy numbers must not be always a (normal) triangular fuzzy number.

2.3.3 *Fuzzy Linear Programming Method of Matrix Games with Payoffs of Triangular Fuzzy Numbers*

According to Definitions 2.4 and 2.5, the maximin (mixed) strategy $\boldsymbol{y}^* \in Y$ and gain-floor $\tilde{\upsilon}^*$ for the player I and the minimax (mixed) strategy $\boldsymbol{z}^* \in Z$ and loss-ceiling $\tilde{\omega}^*$ for the player II can be generated by solving the fuzzy mathematical programming models:

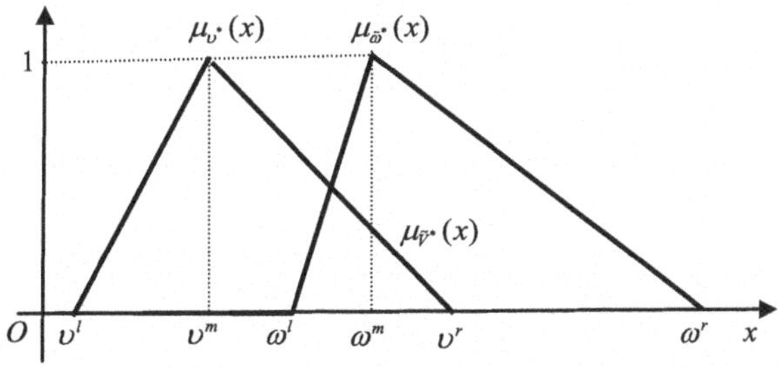

Fig. 2.4 A fuzzy equilibrium value $\tilde{V}^*$

$$\max\{\tilde{\upsilon}\}$$
$$\text{s.t.}\begin{cases} \boldsymbol{y}^{\mathrm{T}}\tilde{\boldsymbol{A}}\boldsymbol{z} \tilde{\geq} \tilde{\upsilon} \quad \text{for all } \boldsymbol{z} \in Z \\ \boldsymbol{y} \in Y \\ \tilde{\upsilon} \in T(\mathrm{R}) \\ \tilde{\upsilon} \text{ unrestricted in sign} \end{cases} \tag{2.6}$$

and

$$\min\{\tilde{\omega}\}$$
$$\text{s.t.}\begin{cases} \boldsymbol{y}^{\mathrm{T}}\tilde{\boldsymbol{A}}\boldsymbol{z} \tilde{\leq} \tilde{\omega} \quad \text{for all } \boldsymbol{y} \in Y \\ \boldsymbol{z} \in Z \\ \tilde{\omega} \in T(\mathrm{R}) \\ \tilde{\omega} \text{ unrestricted in sign,} \end{cases} \tag{2.7}$$

respectively.

It makes sense to consider only the extreme points of the sets Y and Z in the constraints of Eqs. (2.6) and (2.7) since "$\tilde{\leq}$" and "$\tilde{\geq}$" preserve the ranking order when triangular fuzzy numbers are multiplied by positive scalars according to Eq. (2.3) and Definition 2.1. Then, Eqs. (2.6) and (2.7) can be converted into the fuzzy mathematical programming models as follows:

$$\max\{\tilde{\upsilon}\}$$
$$\text{s.t.}\begin{cases} \sum_{i=1}^{m} \tilde{a}_{ij} y_i \tilde{\geq} \tilde{\upsilon} \quad (j = 1, 2, \ldots, n) \\ \sum_{i=1}^{m} y_i = 1 \\ y_i \geq 0 \quad (i = 1, 2, \ldots, m) \\ \tilde{\upsilon} \in T(\mathrm{R}) \\ \tilde{\upsilon} \text{ unrestricted in sign} \end{cases} \tag{2.8}$$

and

$$\min\{\tilde{\omega}\}$$
$$\text{s.t.}\begin{cases} \sum_{j=1}^{n} \tilde{a}_{ij} z_j \tilde{\leq} \tilde{\omega} \quad (i = 1, 2, \ldots, m) \\ \sum_{j=1}^{n} z_j = 1 \\ z_j \geq 0 \quad (j = 1, 2, \ldots, n) \\ \tilde{\omega} \in T(\mathrm{R}) \\ \tilde{\omega} \text{ unrestricted in sign,} \end{cases} \tag{2.9}$$

respectively, where $\tilde{\upsilon}$ and $\tilde{\omega}$ are fuzzy variables, y_i $(i = 1, 2, \ldots, m)$ and z_j $(j = 1, 2, \ldots, n)$ are decision variables.

According to the operations of triangular fuzzy numbers, in general, we can draw an important conclusion, which is summarized as in Theorem 2.2.

Theorem 2.2 *Assume that* $(\mathbf{y}^*, \tilde{\upsilon}^*)$ *and* $(\mathbf{z}^*, \tilde{\omega}^*)$ *are optimal solutions of Eqs.* (2.8) *and* (2.9)*, respectively. Then,* $\tilde{\upsilon}^*$ *and* $\tilde{\omega}^*$ *are triangular fuzzy numbers and* $\tilde{\upsilon}^* \tilde{\leq} \tilde{\omega}^*$.

Proof Due to the assumption that $(\mathbf{y}^*, \tilde{\upsilon}^*)$ and $(\mathbf{z}^*, \tilde{\omega}^*)$ respectively are optimal solutions of Eqs. (2.8) and (2.9), then according to Eqs. (2.2) and (2.3), it follows that $\tilde{\upsilon}^*$ and $\tilde{\omega}^*$ are triangular fuzzy numbers. Furthermore, it follows from Eqs. (2.8) and (2.9) that

$$\begin{aligned} \tilde{\upsilon}^* &= \sum_{j=1}^{n} \tilde{\upsilon}^* z_j^* \tilde{\leq} \sum_{j=1}^{n} \left(\sum_{i=1}^{m} \tilde{a}_{ij} y_i^* \right) z_j^* \\ &= \sum_{i=1}^{m} \left(\sum_{j=1}^{n} \tilde{a}_{ij} z_j^* \right) y_i^* \tilde{\leq} \sum_{i=1}^{m} \tilde{\omega}^* y_i^* = \tilde{\omega}^*, \end{aligned}$$

i.e., $\tilde{\upsilon}^* \tilde{\leq} \tilde{\omega}^*$. Thus, we have finished the proof of Theorem 2.2.

Theorem 2.2 means that the player I's gain-floor "essentially cannot exceed" the player II's loss-ceiling in the sense of Definition 2.1.

Equations (2.8) and (2.9) are general fuzzy mathematical programming models which may involve in different solutions [24, 25]. But in this section, the fuzzy optimization is made in the sense of Definition 2.2 or Definition 2.3. In the following, we will focus on studying the solving method and procedure of Eqs. (2.8) and (2.9).

According to Definitions 2.1–2.3, Eqs. (2.8) and (2.9) can be converted into the multi-objective mathematical programming models as follows:

$$\begin{aligned} &\max\{\upsilon^l\} \\ &\max\{\upsilon^m\} \\ &\max\{\upsilon^r\} \\ &\text{s.t.} \begin{cases} \sum_{i=1}^{m} a_{ij}^l y_i \geq \upsilon^l \quad (j = 1, 2, \ldots, n) \\ \sum_{i=1}^{m} a_{ij}^m y_i \geq \upsilon^m \quad (j = 1, 2, \ldots, n) \\ \sum_{i=1}^{m} a_{ij}^r y_i \geq \upsilon^r \quad (j = 1, 2, \ldots, n) \\ \upsilon^l \leq \upsilon^m \leq \upsilon^r \\ \sum_{i=1}^{m} y_i = 1 \\ y_i \geq 0 \quad (i = 1, 2, \ldots, m) \\ \upsilon^l, \upsilon^m, \text{ and } \upsilon^r \text{ unrestricted in sign} \end{cases} \end{aligned} \tag{2.10}$$

and

$$\begin{aligned}&\min\{\omega^l\}\\&\min\{\omega^m\}\\&\min\{\omega^r\}\\&\text{s.t.}\begin{cases}\sum_{j=1}^{n} a_{ij}^l z_j \leq \omega^l \quad (i=1,2,\ldots,m)\\ \sum_{j=1}^{n} a_{ij}^m z_j \leq \omega^m \quad (i=1,2,\ldots,m)\\ \sum_{j=1}^{n} a_{ij}^r z_j \leq \omega^r \quad (i=1,2,\ldots,m)\\ \omega^l \leq \omega^m \leq \omega^r\\ \sum_{j=1}^{n} z_j = 1\\ z_j \geq 0 \quad (j=1,2,\ldots,n)\\ \omega^l, \omega^m, \text{ and } \omega^r \text{ unrestricted in sign,}\end{cases}\end{aligned} \tag{2.11}$$

respectively.

For the above multi-objective mathematical programming models, there are few standard ways of defining a solution. Normally, the concept of Pareto optimal solutions/efficient solutions is commonly-used [4, 21, 22]. There exist several solution methods for them such as utility theory, goal programming, fuzzy programming, and interactive approaches. However, in the following, we develop a fuzzy linear programming method based on Zimmermann's fuzzy programming method [24] with our normalization process.

Firstly, we can compute the positive ideal solution and negative ideal solution of Eq. (2.10) through solving three linear programming models with different objective functions, respectively. Specifically, using the simplex method of linear programming, we solve the linear programming model as follows:

$$\begin{aligned}&\max\{\upsilon^l\}\\&\text{s.t.}\begin{cases}\sum_{i=1}^{m} a_{ij}^l y_i \geq \upsilon^l \quad (j=1,2,\ldots,n)\\ \sum_{i=1}^{m} a_{ij}^m y_i \geq \upsilon^m \quad (j=1,2,\ldots,n)\\ \sum_{i=1}^{m} a_{ij}^r y_i \geq \upsilon^r \quad (j=1,2,\ldots,n)\\ \upsilon^l \leq \upsilon^m \leq \upsilon^r\\ \sum_{i=1}^{m} y_i = 1\\ y_i \geq 0 \quad (i=1,2,\ldots,m)\\ \upsilon^l, \upsilon^m, \text{ and } \upsilon^r \text{ unrestricted in sign,}\end{cases}\end{aligned}$$

denoted its optimal solution by $(\mathbf{y}^{1+}, \upsilon^{l1+}, \upsilon^{m1+}, \upsilon^{r1+})$.

Analogously, using the simplex method of linear programming, we solve the linear programming model as follows:

$$\max\{\upsilon^m\}$$
$$\text{s.t.}\begin{cases} \sum_{i=1}^{m} a_{ij}^l y_i \geq \upsilon^l \quad (j = 1, 2, \ldots, n) \\ \sum_{i=1}^{m} a_{ij}^m y_i \geq \upsilon^m \quad (j = 1, 2, \ldots, n) \\ \sum_{i=1}^{m} a_{ij}^r y_i \geq \upsilon^r \quad (j = 1, 2, \ldots, n) \\ \upsilon^l \leq \upsilon^m \leq \upsilon^r \\ \sum_{i=1}^{m} y_i = 1 \\ y_i \geq 0 \quad (i = 1, 2, \ldots, m) \\ \upsilon^l, \upsilon^m, \text{ and } \upsilon^r \text{ unrestricted in sign,} \end{cases}$$

denoted its optimal solution by $(\mathbf{y}^{2+}, \upsilon^{l2+}, \upsilon^{m2+}, \upsilon^{r2+})$. We solve the linear programming model as follows:

$$\max\{\upsilon^r\}$$
$$\text{s.t.}\begin{cases} \sum_{i=1}^{m} a_{ij}^l y_i \geq \upsilon^l \quad (j = 1, 2, \ldots, n) \\ \sum_{i=1}^{m} a_{ij}^m y_i \geq \upsilon^m \quad (j = 1, 2, \ldots, n) \\ \sum_{i=1}^{m} a_{ij}^r y_i \geq \upsilon^r \quad (j = 1, 2, \ldots, n) \\ \upsilon^l \leq \upsilon^m \leq \upsilon^r \\ \sum_{i=1}^{m} y_i = 1 \\ y_i \geq 0 \quad (i = 1, 2, \ldots, m) \\ \upsilon^l, \upsilon^m, \text{ and } \upsilon^r \text{ unrestricted in sign,} \end{cases}$$

denoted its optimal solution by $(\mathbf{y}^{3+}, \upsilon^{l3+}, \upsilon^{m3+}, \upsilon^{r3+})$.

Thus, the positive ideal solution of Eq. (2.10) can be obtained as $(\upsilon^{l+}, \upsilon^{m+}, \upsilon^{r+}) = (\upsilon^{l1+}, \upsilon^{m2+}, \upsilon^{r3+})$. The negative ideal solution of Eq. (2.10) can be defined as follows:

$$\begin{aligned}(\upsilon^{l-}, \upsilon^{m-}, \upsilon^{r-}) = (&\min\{\upsilon^{lt+} | t = 1, 2, 3\}, \\ &\min\{\upsilon^{mt+} | t = 1, 2, 3\}, \min\{\upsilon^{r3+} | t = 1, 2, 3\}).\end{aligned}$$

Hereby, the relative membership functions of the three objective functions in Eq. (2.10) can be defined as follows:

$$\eta_{\upsilon^l}(\upsilon^l) = \begin{cases} 1 & \text{if } \upsilon^l \geq \upsilon^{l+} \\ \dfrac{\upsilon^l - \upsilon^{l-}}{\upsilon^{l+} - \upsilon^{l-}} & \text{if } \upsilon^{l-} \leq \upsilon^l < \upsilon^{l+} \\ 0 & \text{if } \upsilon^l < \upsilon^{l-}, \end{cases}$$

$$\eta_{\upsilon^m}(\upsilon^m) = \begin{cases} 1 & \text{if } \upsilon^m \geq \upsilon^{m+} \\ \dfrac{\upsilon^m - \upsilon^{m-}}{\upsilon^{m+} - \upsilon^{m-}} & \text{if } \upsilon^{m-} \leq \upsilon^m < \upsilon^{m+} \\ 0 & \text{if } \upsilon^m < \upsilon^{m-} \end{cases}$$

and

$$\eta_{\upsilon^r}(\upsilon^r) = \begin{cases} 1 & \text{if } \upsilon^r \geq \upsilon^{r+} \\ \dfrac{\upsilon^r - \upsilon^{r-}}{\upsilon^{r+} - \upsilon^{r-}} & \text{if } \upsilon^{r-} \leq \upsilon^r < \upsilon^{r+} \\ 0 & \text{if } \upsilon^r < \upsilon^{r-}, \end{cases}$$

respectively.

Using Zimmermann's fuzzy programming method [24], Eq. (2.10) can be converted into the linear programming model as follows:

$$\max\{\eta\}$$
$$\text{s.t.} \begin{cases} \sum_{i=1}^{m} a_{ij}^l y_i \geq \upsilon^l \quad (j = 1, 2, \ldots, n) \\ \sum_{i=1}^{m} a_{ij}^m y_i \geq \upsilon^m \quad (j = 1, 2, \ldots, n) \\ \sum_{i=1}^{m} a_{ij}^r y_i \geq \upsilon^r \quad (j = 1, 2, \ldots, n) \\ \upsilon^l - \upsilon^{l-} \geq (\upsilon^{l+} - \upsilon^{l-})\eta \\ \upsilon^m - \upsilon^{m-} \geq (\upsilon^{m+} - \upsilon^{m-})\eta \\ \upsilon^r - \upsilon^{r-} \geq (\upsilon^{r+} - \upsilon^{r-})\eta \\ \upsilon^l \leq \upsilon^m \leq \upsilon^r \\ \sum_{i=1}^{m} y_i = 1 \\ 0 \leq \eta \leq 1 \\ y_i \geq 0 \quad (i = 1, 2, \ldots, m) \\ \upsilon^l, \upsilon^m, \text{ and } \upsilon^r \text{ unrestricted in sign,} \end{cases} \tag{2.12}$$

where $\eta = \min\{\eta_{\upsilon^l}(\upsilon^l), \eta_{\upsilon^m}(\upsilon^m), \eta_{\upsilon^r}(\upsilon^r)\}$.

Solving Eq. (2.12) by using the simplex method of linear programming, we can obtain the optimal or maximin (mixed) strategy $\boldsymbol{y}^*$ and gain-floor $\tilde{\upsilon}^*$ for the player I.

In the same way to the above consideration of Eq. (2.10), according to Eq. (2.11), using the simplex method of linear programming, we can solve the linear programming model as follows:

$$\min\{\omega^l\}$$
$$\text{s.t.}\begin{cases} \sum_{j=1}^{n} a_{ij}^l z_j \leq \omega^l \quad (i = 1, 2, \ldots, m) \\ \sum_{j=1}^{n} a_{ij}^m z_j \leq \omega^m \quad (i = 1, 2, \ldots, m) \\ \sum_{j=1}^{n} a_{ij}^r z_j \leq \omega^r \quad (i = 1, 2, \ldots, m) \\ \omega^l \leq \omega^m \leq \omega^r \\ \sum_{j=1}^{n} z_j = 1 \\ z_j \geq 0 \quad (j = 1, 2, \ldots, n) \\ \omega^l, \omega^m, \text{ and } \omega^r \text{ unrestricted in sign,} \end{cases}$$

denoted its optimal solution by $(z^{1+}, \omega^{l1+}, \omega^{m1+}, \omega^{r1+})$. Analogously, we can solve the linear programming model as follows:

$$\min\{\omega^m\}$$
$$\text{s.t.}\begin{cases} \sum_{j=1}^{n} a_{ij}^l z_j \leq \omega^l \quad (i = 1, 2, \ldots, m) \\ \sum_{j=1}^{n} a_{ij}^m z_j \leq \omega^m \quad (i = 1, 2, \ldots, m) \\ \sum_{j=1}^{n} a_{ij}^r z_j \leq \omega^r \quad (i = 1, 2, \ldots, m) \\ \omega^l \leq \omega^m \leq \omega^r \\ \sum_{j=1}^{n} z_j = 1 \\ z_j \geq 0 \quad (j = 1, 2, \ldots, n) \\ \omega^l, \omega^m, \text{ and } \omega^r \text{ unrestricted in sign,} \end{cases}$$

denoted its optimal solution by $(z^{2+}, \omega^{l2+}, \omega^{m2+}, \omega^{r2+})$. We can solve the linear programming model as follows:

$$\min\{\omega^r\}$$
$$\text{s.t.}\begin{cases} \sum_{j=1}^{n} a_{ij}^l z_j \leq \omega^l \quad (i = 1, 2, \ldots, m) \\ \sum_{j=1}^{n} a_{ij}^m z_j \leq \omega^m \quad (i = 1, 2, \ldots, m) \\ \sum_{j=1}^{n} a_{ij}^r z_j \leq \omega^r \quad (i = 1, 2, \ldots, m) \\ \omega^l \leq \omega^m \leq \omega^r \\ \sum_{j=1}^{n} z_j = 1 \\ z_j \geq 0 \quad (j = 1, 2, \ldots, n) \\ \omega^l, \omega^m, \text{ and } \omega^r \text{ unrestricted in sign,} \end{cases}$$

denoted its optimal solution by $(z^{3+}, \omega^{l3+}, \omega^{m3+}, \omega^{r3+})$.

Then, the positive ideal solution of Eq. (2.11) can be obtained as $(\omega^{l+}, \omega^{m+}, \omega^{r+}) = (\omega^{l1+}, \omega^{m2+}, \omega^{r3+})$. The negative ideal solution of Eq. (2.11) can be defined as follows:

$$(\omega^{l-}, \omega^{m-}, \omega^{r-}) = (\max\{\omega^{lt+} | t = 1, 2, 3\}, \\ \max\{\omega^{mt+} | t = 1, 2, 3\}, \max\{\omega^{r3+} | t = 1, 2, 3\}).$$

Hereby, the relative membership functions of the three objective functions in Eq. (2.11) can be defined as follows:

$$\rho_{\omega^l}(\omega^l) = \begin{cases} 1 & \text{if } \omega^l \leq \omega^{l+} \\ \dfrac{\omega^l - \omega^{l+}}{\omega^{l-} - \omega^{l+}} & \text{if } \omega^{l+} < \omega^l \leq \omega^{l-} \\ 0 & \text{if } \omega^l > \omega^{l-}, \end{cases}$$

$$\rho_{\omega^m}(\omega^m) = \begin{cases} 1 & \text{if } \omega^m \leq \omega^{m+} \\ \dfrac{\omega^m - \omega^{m+}}{\omega^{m-} - \omega^{m+}} & \text{if } \omega^{m+} < \omega^m \leq \omega^{m-} \\ 0 & \text{if } \omega^m > \omega^{m-} \end{cases}$$

and

$$\rho_{\omega^r}(\omega^r) = \begin{cases} 1 & \text{if } \quad \omega^r \leq \omega^{r+} \\ \dfrac{\omega^r - \omega^{r+}}{\omega^{r-} - \omega^{r+}} & \text{if } \quad \omega^{r+} < \omega^r \leq \omega^{r-} \\ 0 & \text{if } \quad \omega^r > \omega^{r-}, \end{cases}$$

respectively.

Using Zimmermann's fuzzy programming method [24], Eq. (2.11) can be converted into the linear programming model as follows:

$$\max\{\rho\}$$
$$\text{s.t.}\begin{cases} \sum_{j=1}^{n} a_{ij}^{l} z_j \le \omega^{l} \quad (i=1,2,\ldots,m) \\ \sum_{j=1}^{n} a_{ij}^{m} z_j \le \omega^{m} \quad (i=1,2,\ldots,m) \\ \sum_{j=1}^{n} a_{ij}^{r} z_j \le \omega^{r} \quad (i=1,2,\ldots,m) \\ \omega^{l}-\omega^{l+} \ge (\omega^{l-}-\omega^{l+})\rho \\ \omega^{m}-\omega^{m+} \ge (\omega^{m-}-\omega^{m+})\rho \\ \omega^{r}-\omega^{r+} \ge (\omega^{r-}-\omega^{r+})\rho \\ \omega^{l} \le \omega^{m} \le \omega^{r} \\ \sum_{j=1}^{n} z_j = 1 \\ 0 \le \rho \le 1 \\ z_j \ge 0 \quad (j=1,2,\ldots,n) \\ \omega^{l}, \omega^{m}, \text{ and } \omega^{r} \text{ unrestricted in sign,} \end{cases} \tag{2.13}$$

where $\rho = \min\{\rho_{\omega^l}(\omega^l), \rho_{\omega^m}(\omega^m), \rho_{\omega^r}(\omega^r)\}$.

Solving Eq. (2.13) by using the simplex method of linear programming, we can obtain the optimal or minimax (mixed) strategy z^* and loss-ceiling $\tilde{\omega}^*$ for the player II.

Example 2.1 Let us consider a simple numerical example of matrix games with payoffs of triangular fuzzy numbers. Assume that the payoff matrix for the player I is given as follows:

$$\tilde{\boldsymbol{A}}_1 = \begin{matrix} \\ \delta_1 \\ \delta_2 \end{matrix} \begin{matrix} \beta_1 & \beta_2 \\ \left(\begin{matrix} (18,20,23) \\ (-33,-32,-27) \end{matrix} \right. & \left. \begin{matrix} (-21,-18,-16) \\ (38,40,43) \end{matrix} \right) \end{matrix}.$$

According to Eqs. (2.12) and (2.13), we can construct two linear programming models for the players I and II, respectively. Using the simplex method of linear programming, we can easily obtain their optimal solutions whose components are given as follows:

$$\boldsymbol{y}_1^* = (0.648, 0.352)^{\mathrm{T}},$$
$$\tilde{v}_1^* = (-0.254, 1.715, 4.746),$$
$$\eta_1^* = 0.501,$$
$$\boldsymbol{z}_1^* = (0.534, 0.466)^{\mathrm{T}},$$
$$\tilde{\omega}_1^* = (0.241, 2.303, 5.601)$$

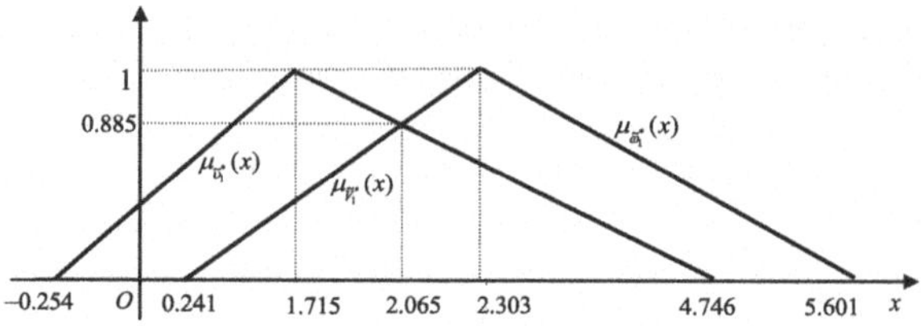

Fig. 2.5 The fuzzy equilibrium value $\tilde{V}_1^*$

and

$$\rho_1^* = 0.500,$$

respectively. Furthermore, we have

$$\mu_{\tilde{V}_1^*}(x) = \begin{cases} \dfrac{x - 0.241}{2.062} & \text{if } 0.241 \le x < 2.065 \\ 0.885 & \text{if } x = 2.065 \\ \dfrac{4.746 - x}{3.031} & \text{if } 2.065 < x \le 4.746 \\ 0 & \text{else.} \end{cases}$$

Therefore, there exists a fuzzy equilibrium value 2.065 with the possibility of 0.885. In other words, the fuzzy value of the matrix game $\tilde{\boldsymbol{A}}_1$ with payoffs of triangular fuzzy numbers is "around 2.065". Or the player I's minimum reward is 0.241 while his/her maximum reward is 4.746. The player I can win any intermediate value x between 0.241 and 4.746 with the possibility $\mu_{\tilde{V}_1^*}(x)$, depicted as in Fig. 2.5.

2.4 Two-Level Linear Programming Models of Matrix Games with Payoffs of Triangular Fuzzy Numbers

Stated as in Sect. 2.3, Eqs. (2.10) and (2.11) are multi-objective linear programming models, which may be solved by several methods [21, 22]. However, in this section, we develop a two-level linear programming method for solving Eqs. (2.10) and (2.11).

In Eq. (2.10), the three objective functions (i.e., υ^l, υ^m, and υ^r) should have different priority. In fact, the objective functions may be written as the triangular fuzzy number $\tilde{\upsilon} = (\upsilon^l, \upsilon^m, \upsilon^r)$, where υ^m is the mean (or center) of the triangular fuzzy number $\tilde{\upsilon}$, and υ^l and υ^r are lower and upper limits (or bounds) of the triangular fuzzy number $\tilde{\upsilon}$, respectively. The priority of the objective function υ^m

should be higher than that of both the objective functions υ^l and υ^r, and the priority of υ^l and υ^r may be identical because the priority of the mean of the triangular fuzzy number is much higher than that of its lower and upper limits according to the fuzzy sets [3, 4, 24]. Hence, Eq. (2.10) may be regarded as a two-level linear programming problem. Its first priority is given to the objective function υ^m. Its second priority is given to the objective functions υ^l and υ^r. Thus, solving Eq. (2.10) becomes solving the following linear programming models [i.e., Eqs. (2.14) and (2.15)] successively. To be more specific, we give its procedure as follows.

According to Eq. (2.10), the linear programming model in the first level is constructed as follows:

$$\max\{\upsilon^m\}$$
$$\text{s.t.}\begin{cases} \sum_{i=1}^{m} a_{ij}^l y_i \geq \upsilon^l & (j=1,2,\ldots,n) \\ \sum_{i=1}^{m} a_{ij}^m y_i \geq \upsilon^m & (j=1,2,\ldots,n) \\ \sum_{i=1}^{m} a_{ij}^r y_i \geq \upsilon^r & (j=1,2,\ldots,n) \\ \upsilon^l \leq \upsilon^m \leq \upsilon^r \\ \sum_{i=1}^{m} y_i = 1 \\ y_i \geq 0 \quad (i=1,2,\ldots,m) \\ \upsilon^l, \upsilon^m, \text{ and } \upsilon^r \text{ unrestricted in sign,} \end{cases} \tag{2.14}$$

where y_i $(i = 1,2,\ldots,m)$, υ^l, υ^m, and υ^r are decision variables. Using the simplex method of linear programming, we can obtain its optimal solution by $(\boldsymbol{y}^*, \upsilon^{l0}, \upsilon^{m*}, \upsilon^{r0})$, where $\boldsymbol{y}^* = (y_1^*, y_2^*, \ldots, y_m^*)^{\mathrm{T}}$.

Combining with Eq. (2.10), the linear programming model in the second level is constructed as follows:

$$\max\{\upsilon^l\}$$
$$\max\{\upsilon^r\}$$
$$\text{s.t.}\begin{cases} \sum_{i=1}^{m} a_{ij}^l y_i^* \geq \upsilon^l & (j=1,2,\ldots,n) \\ \sum_{i=1}^{m} a_{ij}^r y_i^* \geq \upsilon^r & (j=1,2,\ldots,n) \\ \upsilon^l \geq \upsilon^{l0} \\ \upsilon^r \geq \upsilon^{r0} \\ \upsilon^l \text{ and } \upsilon^r \text{ unrestricted in sign,} \end{cases} \tag{2.15}$$

where υ^l and υ^r are decision variables.

In Eq. (2.15), adding the constraints $\upsilon^l \geq \upsilon^{l0}$ and $\upsilon^r \geq \upsilon^{r0}$ aim to improve the objective functions υ^l and υ^r, respectively. It is the real reason why the second-level linear programming model [i.e., Eq. (2.15)] is introduced after the first-level linear programming model [i.e., Eq. (2.14)].

It is easy to see from Eq. (2.15) that the constraints of the variable υ^l are independent of those of the variable υ^r. Therefore, Eq. (2.15) can be decompounded into the two linear programming models as follows:

$$\max\{\upsilon^l\}$$
$$\text{s.t.}\begin{cases} \sum_{i=1}^{m} a_{ij}^l y_i^* \geq \upsilon^l \quad (j = 1, 2, \ldots, n) \\ \upsilon^l \geq \upsilon^{l0} \\ \upsilon^l \text{ unrestricted in sign} \end{cases} \tag{2.16}$$

and

$$\max\{\upsilon^r\}$$
$$\text{s.t.}\begin{cases} \sum_{i=1}^{m} a_{ij}^l y_i^* \geq \upsilon^r \quad (j = 1, 2, \ldots, n) \\ \upsilon^r \geq \upsilon^{r0} \\ \upsilon^r \text{ unrestricted in sign.} \end{cases} \tag{2.17}$$

Solving Eqs. (2.16) and (2.17) by using the simplex method of linear programming, we can obtain their optimal solutions υ^{l*} and υ^{r*}, respectively.

It is not difficult to prove that $(\boldsymbol{y}^*, \tilde{\upsilon}^*)$ is a Pareto optimal solution of Eq. (2.10), where $\tilde{\upsilon}^* = (\upsilon^{l*}, \upsilon^{m*}, \upsilon^{r*})$ is a triangular fuzzy number. Thus, the optimal (or maximin) mixed strategy $\boldsymbol{y}^*$ and the gain-floor $\tilde{\upsilon}^*$ for the player I can be obtained.

In the same way to the above consideration of Eq. (2.10), the three objective functions ω^l, ω^m, and ω^r of Eq. (2.11) should have different priority. Namely, the priority of the objective function ω^m should be higher than that of both the objective functions ω^l, and ω^r, and the priority of ω^l and ω^r should be assumed to be identical in that ω^m, ω^l, and ω^r are the mean and the lower and upper limits of the triangular fuzzy number $\tilde{\omega} = (\omega^l, \omega^m, \omega^r)$, respectively. Thus, Eq. (2.11) may be regarded as a two-level linear programming problem. Its first priority is given to the objective function ω^m. Its second priority is given to the objective functions ω^l and ω^r. As a result, solving Eq. (2.11) turns into solving the following two linear programming models [i.e., Eqs. (2.18) and (2.19)] successively.

According to Eq. (2.11), the linear programming model in the first level is constructed as follows:

$$\min\{\omega^m\}$$
$$\text{s.t.}\begin{cases} \sum_{j=1}^{n} a_{ij}^l z_j \le \omega^l \quad (i = 1, 2, \ldots, m) \\ \sum_{j=1}^{n} a_{ij}^m z_j \le \omega^m \quad (i = 1, 2, \ldots, m) \\ \sum_{j=1}^{n} a_{ij}^r z_j \le \omega^r \quad (i = 1, 2, \ldots, m) \\ \omega^l \le \omega^m \le \omega^r \\ \sum_{j=1}^{n} z_j = 1 \\ z_j \ge 0 \quad (j = 1, 2, \ldots, n) \\ \omega^l, \omega^m, \text{ and } \omega^r \text{ unrestricted in sign,} \end{cases} \tag{2.18}$$

where z_j $(j = 1, 2, \ldots, n)$, ω^l, ω^m, and ω^r are decision variables. Solving Eq. (2.18) by using the simplex method of linear programming, we can easily obtain its optimal solution $(z^*, \omega^{l0}, \omega^{m*}, \omega^{r0})$, where $z^* = (z_1^*, z_2^*, \ldots, z_n^*)^{\mathrm{T}}$.

Combining with Eq. (2.11), the linear programming model in the second level is constructed as follows:

$$\min\{\omega^l\}$$
$$\min\{\omega^r\}$$
$$\text{s.t.}\begin{cases} \sum_{j=1}^{n} a_{ij}^l z_j^* \le \omega^l \quad (i = 1, 2, \ldots, m) \\ \sum_{j=1}^{n} a_{ij}^r z_j^* \le \omega^r \quad (i = 1, 2, \ldots, m) \\ \omega^l \le \omega^{l0} \\ \omega^r \le \omega^{r0} \\ \omega^l \text{ and } \omega^r \text{ unrestricted in sign,} \end{cases} \tag{2.19}$$

where ω^l and ω^r are decision variables.

Analogously, adding the constraints $\omega^l \le \omega^{l0}$ and $\omega^r \le \omega^{r0}$ in Eq. (2.19) aim to improve ω^l and ω^r, respectively.

It is easy to see from Eq. (2.19) that the constraints of the variable ω^l are independent of those of the variable ω^r. Therefore, Eq. (2.19) can be decompounded into the linear programming models as follows:

$$\min\{\omega^l\}$$
$$\text{s.t.}\begin{cases}\sum_{j=1}^{n} a_{ij}^l z_j^* \le \omega^l \quad (i=1,2,\ldots,m)\\ \omega^l \le \omega^{l0}\\ \omega^l \ \text{unrestricted in sign}\end{cases} \tag{2.20}$$

and

$$\min\{\omega^r\}$$
$$\text{s.t.}\begin{cases}\sum_{j=1}^{n} a_{ij}^r z_j^* \le \omega^r \quad (i=1,2,\cdots,m)\\ \omega^r \le \omega^{r0}\\ \omega^r \ \text{unrestricted in sign.}\end{cases} \tag{2.21}$$

Solving Eqs. (2.20) and (2.21) through using the simplex method of linear programming, we can easily obtain their solutions ω^{l*} and ω^{r*}, respectively.

It is not difficult to prove that $(z^*, \tilde{\omega}^*)$ is a Pareto optimal solution of Eq. (2.11), where $\tilde{\omega}^* = (\omega^{l*}, \omega^{m*}, \omega^{r*})$ is a triangular fuzzy number. Thus, the optimal (or minimax) mixed strategy z^* and the loss-ceiling $\tilde{\omega}^*$ for the player II can be obtained.

Hence, $(\boldsymbol{y}^*, \boldsymbol{z}^*, \tilde{\upsilon}^*, \tilde{\omega}^*)^{\mathrm{T}}$ and $\tilde{V}^* = \tilde{\upsilon}^* \wedge \tilde{\omega}^*$ are a solution and a fuzzy equilibrium value of the matrix game $\tilde{\boldsymbol{A}}$ with payoffs of triangular fuzzy numbers, respectively.

Example 2.2 Let us consider a simple numerical example which is taken from Campos [7]. Suppose that the payoff matrix for the player I is given as follows:

$$\tilde{\boldsymbol{A}}_2 = \begin{matrix} \\ \delta_1 \\ \delta_2 \end{matrix}\begin{matrix} \beta_1 & \beta_2 \\ \left(\begin{matrix}(175,180,190) \\ (80,90,100)\end{matrix}\right. & \left.\begin{matrix}(150,156,158) \\ (175,180,190)\end{matrix}\right)\end{matrix},$$

where all elements of the above payoff matrix $\tilde{\boldsymbol{A}}_2$ are triangular fuzzy numbers.

According to Eq. (2.14), the linear programming model in the first level can be constructed as follows:

$$\max\{\upsilon^m\}$$
$$\text{s.t.}\begin{cases}175y_1 + 80y_2 \ge \upsilon^l\\ 150y_1 + 175y_2 \ge \upsilon^l\\ 180y_1 + 90y_2 \ge \upsilon^m\\ 156y_1 + 180y_2 \ge \upsilon^m\\ 190y_1 + 100y_2 \ge \upsilon^r\\ 158y_1 + 190y_2 \ge \upsilon^r\\ \upsilon^l \le \upsilon^m \le \upsilon^r\\ y_1 + y_2 = 1\\ y_1 \ge 0, y_2 \ge 0\\ \upsilon^l, \upsilon^m, \ \text{and} \ \upsilon^r \ \text{unrestricted in sign.}\end{cases}$$

Solving the above linear programming model by using the simplex method of linear programming, we can obtain its optimal solution $(\boldsymbol{y}^*, \upsilon^{l0}, \upsilon^{m*}, \upsilon^{r0})$, where $\boldsymbol{y}^* = (0.7895, 0.2105)^{\mathrm{T}}$, $\upsilon^{l0} = 61.398$, $\upsilon^{m*} = 161.05$, and $\upsilon^{r0} = 163.063$.

According to Eqs. (2.16) and (2.17), the two linear programming models in the second level can be constructed as follows:

$$\max\{\upsilon^l\}$$
$$\text{s.t.}\begin{cases} \upsilon^l \le 154.9996 \\ \upsilon^l \le 155.2633 \\ \upsilon^l \ge 61.398 \\ \upsilon^l \text{ unrestricted in sign} \end{cases}$$

and

$$\max\{\upsilon^r\}$$
$$\text{s.t.}\begin{cases} \upsilon^r \le 171.0523 \\ \upsilon^r \le 164.737 \\ \upsilon^r \ge 163.063 \\ \upsilon^r \text{ unrestricted in sign,} \end{cases}$$

respectively. It is easy to see that $\upsilon^{l*} = 154.9996$ and $\upsilon^{r*} = 164.737$ are the solutions of the above two linear programming models, respectively.

Therefore, the optimal (or maximin) mixed strategy and the gain-floor for the player I are $\boldsymbol{y}^* = (0.7895, 0.2105)^{\mathrm{T}}$ and $\tilde{\upsilon}^* = (154.9996, 161.05, 164.737)$, respectively.

Analogously, according to Eq. (2.18), the linear programming model in the first level can be constructed as follows:

$$\min\{\omega^m\}$$
$$\text{s.t.}\begin{cases} 175z_1 + 150z_2 \le \omega^l \\ 80z_1 + 175z_2 \le \omega^l \\ 180z_1 + 156z_2 \le \omega^m \\ 90z_1 + 180z_2 \le \omega^m \\ 190z_1 + 158z_2 \le \omega^r \\ 100z_1 + 190z_2 \le \omega^r \\ \omega^l \le \omega^m \le \omega^r \\ z_1 + z_2 = 1 \\ z_1 \ge 0, z_2 \ge 0 \\ \omega^l, \omega^m, \text{ and } \omega^r \text{ unrestricted in sign.} \end{cases}$$

Solving the above linear programming model by using the simplex method of linear programming, we can obtain its optimal solution $(z^*, \omega^{l0}, \omega^{m*}, \omega^{r0})$, where $z^* = (0.2105, 0.7895)^{\mathrm{T}}$, $\omega^{l0} = 158.8984$, $\omega^{m*} = 161.05$, and $\omega^{r0} = 339.61$.

According to Eqs. (2.20) and (2.21), the two linear programming models in the second level can be constructed as follows:

$$\min\{\omega^l\}$$
$$\text{s.t.} \begin{cases} \omega^l \geq 155.2633 \\ \omega^l \geq 154.9997 \\ \omega^l \leq 158.8984 \\ \omega^l \text{ unrestricted in sign} \end{cases}$$

and

$$\min\{\omega^r\}$$
$$\text{s.t.} \begin{cases} \omega^r \geq 164.737 \\ \omega^r \geq 171.0523 \\ \omega^r \leq 339.61 \\ \omega^r \text{ unrestricted in sign,} \end{cases}$$

respectively. It is easy to see that $\omega^{l*} = 155.2633$ and $\omega^{r*} = 171.0523$ are the solutions of the above linear programming models, respectively.

Thus, the optimal (or minimax) mixed strategy and the loss-ceiling for the player II are obtained as $z^* = (0.2105, 0.7895)^{\mathrm{T}}$ and $\tilde{\omega}^* = (155.2633, 161.05, 171.0523)$, respectively. Furthermore, we can obtain the fuzzy equilibrium value of the matrix game $\tilde{\boldsymbol{A}}_2$ with payoffs of triangular fuzzy numbers as follows:

$$\tilde{V}^* = \tilde{\upsilon}^* \wedge \tilde{\omega}^* = (155.2633, 161.05, 164.737),$$

which means that the fuzzy value of the matrix game $\tilde{\boldsymbol{A}}_2$ with payoffs of triangular fuzzy numbers is "around 161.05". In other words, the player I's minimum reward is 155.2633 while his/her maximum reward is 164.737. He/she could win any intermediate value x between 155.2633 and 164.737 with the possibility $\mu_{\tilde{V}^*}(x)$ as follows:

$$\mu_{\tilde{V}^*}(x) = \begin{cases} \dfrac{x - 155.2633}{5.7867} & \text{if } 155.2633 \leq x < 161.05 \\ 1 & \text{if } x = 161.05 \\ \dfrac{164.737 - x}{3.687} & \text{if } 161.05 < x \leq 164.737 \\ 0 & \text{else,} \end{cases}$$

depicted as in Fig. 2.6.

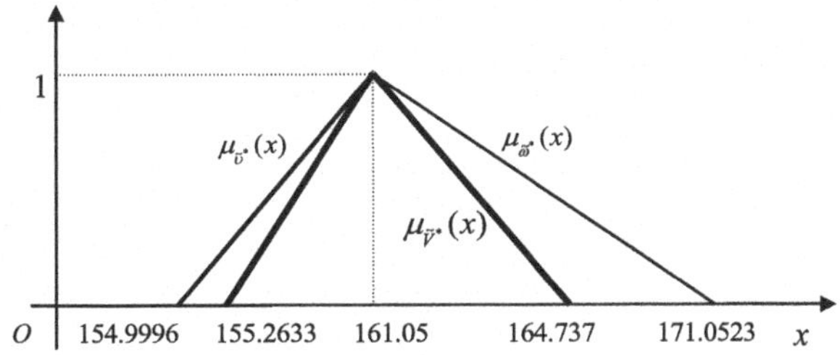

Fig. 2.6 The fuzzy equilibrium value $\tilde{V}^*$

It is easy to see from Fig. 2.6 that the fuzzy equilibrium value $\tilde{V}^*$ is a triangular fuzzy number.

Campos [7] solved the above matrix game $\tilde{\boldsymbol{A}}_2$ with payoffs of triangular fuzzy numbers by deriving two auxiliary fuzzy linear programming models according to four different kinds of ranking methods for fuzzy numbers, and obtained its four fuzzy values and optimal mixed strategies, respectively. The optimal mixed strategies for both the players provided by Campos [7] are almost the same as that generated by using the two-level linear programming method proposed in this section. However, the ranking method for fuzzy numbers needs to be determined a priori, when the method proposed by Campos [7] is employed to solve the matrix game $\tilde{\boldsymbol{A}}_2$ with payoffs of triangular fuzzy numbers. Obviously, it is difficult for the players to determine what kind of ranking methods should be chosen. Moreover, the fuzzy values generated by using the method proposed by Campos [7] closely depend on some additional parameters which are not easy to be chosen for the players.

2.5 The Lexicographic Method of Matrix Games with Payoffs of Triangular Fuzzy Numbers

Let us continue to develop an effective method for solving Eqs. (2.10) and (2.11) stated as in Sect. 2.3.

As stated in Sect. 2.4, the three objective functions υ^l, υ^m, and υ^r in Eq. (2.10) have different priority. Consequently, solving Eq. (2.10) becomes solving the following linear programming problem which consists of the two linear programming models [i.e., Eqs. (2.14) and (2.22)].

Firstly, we solve Eq. (2.14) by using the simplex method of linear programming and obtain its optimal solution, denoted by $(\boldsymbol{y}^0, \upsilon^{l0}, \upsilon^{m*}, \upsilon^{r0})$, where $\boldsymbol{y}^0 = (y_1^0, y_2^0, \ldots, y_m^0)^{\mathrm{T}}$.

Then, combining with Eq. (2.10), the bi-objective linear programming model is constructed as follows:

$$\begin{aligned}
&\max\{\upsilon^l\}\\
&\max\{\upsilon^r\}\\
&\text{s.t.}\begin{cases}
\sum_{i=1}^{m} a_{ij}^l y_i \geq \upsilon^l \quad (j=1,2,\ldots,n)\\
\sum_{i=1}^{m} a_{ij}^m y_i \geq \upsilon^{m*} \quad (j=1,2,\ldots,n)\\
\sum_{i=1}^{m} a_{ij}^r y_i \geq \upsilon^r \quad (j=1,2,\ldots,n)\\
\upsilon^l \leq \upsilon^{m*} \leq \upsilon^r\\
\upsilon^l \geq \upsilon^{l0}\\
\upsilon^r \geq \upsilon^{r0}\\
\sum_{i=1}^{m} y_i = 1\\
y_i \geq 0 \quad (i=1,2,\ldots,m)\\
\upsilon^l \text{ and } \upsilon^r \text{ unrestricted in sign,}
\end{cases}
\end{aligned} \tag{2.22}$$

where y_i $(i=1,2,\ldots,m)$, υ^l, and υ^r are decision variables.

The objective functions υ^l and υ^r in Eq. (2.22) may be regarded as equal importance, i.e., they have identical weights. Therefore, Eq. (2.22) can be aggregated into the linear programming model as follows:

$$\begin{aligned}
&\max\left\{\frac{\upsilon^l+\upsilon^r}{2}\right\}\\
&\text{s.t.}\begin{cases}
\sum_{i=1}^{m} a_{ij}^l y_i \geq \upsilon^l \quad (j=1,2,\ldots,n)\\
\sum_{i=1}^{m} a_{ij}^m y_i \geq \upsilon^{m*} \quad (j=1,2,\ldots,n)\\
\sum_{i=1}^{m} a_{ij}^r y_i \geq \upsilon^r \quad (j=1,2,\ldots,n)\\
\upsilon^l \leq \upsilon^{m*} \leq \upsilon^r\\
\upsilon^l \geq \upsilon^{l0}\\
\upsilon^r \geq \upsilon^{r0}\\
\sum_{i=1}^{m} y_i = 1\\
y_i \geq 0 \quad (i=1,2,\ldots,m)\\
\upsilon^l \text{ and } \upsilon^r \text{ unrestricted in sign.}
\end{cases}
\end{aligned} \tag{2.23}$$

Using the simplex method of linear programming, we can obtain the optimal solution of Eq. (2.23), denoted by $(\boldsymbol{y}^*, \upsilon^{l*}, \upsilon^{r*})$, where $\boldsymbol{y}^* = (y_1^*, y_2^*, \ldots, y_m^*)^{\mathrm{T}}$.

It is not difficult to prove that $(\boldsymbol{y}^*, \tilde{\upsilon}^*)$ is a Pareto optimal solution of Eq. (2.10), where $\tilde{\upsilon}^* = (\upsilon^{l*}, \upsilon^{m*}, \upsilon^{r*})$ is a triangular fuzzy number. Thus, the maximin (or optimal) mixed strategy $\boldsymbol{y}^*$ and the gain-floor $\tilde{\upsilon}^*$ for the player I can be obtained.

In the similar way, solving Eq. (2.11) turns into solving the following linear programming problem which consists of Eqs. (2.18) and (2.24).

Solving Eq. (2.18) by using the simplex method of linear programming, we can easily obtain its optimal solution $(\boldsymbol{z}^0, \omega^{l0}, \omega^{m*}, \omega^{r0})$, where $\boldsymbol{z}^0 = (z_1^0, z_2^0, \ldots, z_n^0)^{\mathrm{T}}$.

Combining with Eq. (2.16), the bi-objective linear programming model is constructed as follows:

$$\begin{aligned} &\min\{\omega^l\} \\ &\min\{\omega^r\} \\ &\text{s.t.}\begin{cases} \sum_{j=1}^{n} a_{ij}^l z_j \le \omega^l \quad (i = 1, 2, \ldots, m) \\ \sum_{j=1}^{n} a_{ij}^m z_j \le \omega^{m*} \quad (i = 1, 2, \ldots, m) \\ \sum_{j=1}^{n} a_{ij}^r z_j \le \omega^r \quad (i = 1, 2, \ldots, m) \\ \omega^l \le \omega^{m*} \le \omega^r \\ \omega^l \le \omega^{l0} \\ \omega^r \le \omega^{r0} \\ \sum_{j=1}^{n} z_j = 1 \\ z_j \ge 0 \quad (j = 1, 2, \ldots, n) \\ \omega^l \text{ and } \omega^r \text{ unrestricted in sign,} \end{cases} \end{aligned} \tag{2.24}$$

where z_j $(j = 1, 2, \ldots, n)$, ω^l, and ω^r are decision variables.

Analogously, the objective functions ω^l and ω^r in Eq. (2.24) may be regarded as equal importance, i.e., they have identical weights. Then, Eq. (2.24) can be aggregated into the linear programming model as follows:

$$
\min\left\{\frac{\omega^l+\omega^r}{2}\right\}
$$
$$
\text{s.t.}\begin{cases}
\sum_{j=1}^{n} a_{ij}^{l} z_j \le \omega^l \quad (i=1,2,\ldots,m)\\
\sum_{j=1}^{n} a_{ij}^{m} z_j \le \omega^{m*} \quad (i=1,2,\ldots,m)\\
\sum_{j=1}^{n} a_{ij}^{r} z_j \le \omega^r \quad (i=1,2,\ldots,m)\\
\omega^l \le \omega^{m*} \le \omega^r\\
\omega^l \le \omega^{l0}\\
\omega^r \le \omega^{r0}\\
\sum_{j=1}^{n} z_j = 1\\
z_j \ge 0 \quad (j=1,2,\ldots,n)\\
\omega^l \text{ and } \omega^r \text{ unrestricted in sign.}
\end{cases}
\tag{2.25}
$$

Solving Eq. (2.25) by using the simplex method of linear programming, we can easily obtain its optimal solution $(z^*, \omega^{l*}, \omega^{r*})$, where $z^* = (z_1^*, z_2^*, \ldots, z_n^*)^{\mathrm{T}}$.

It is easily proved that $(z^*, \tilde{\omega}^*)$ is a Pareto optimal solution of Eq. (2.11), where $\tilde{\omega}^* = (\omega^{l*}, \omega^{m*}, \omega^{r*})$ is a triangular fuzzy number. Thus, the minimax (or optimal) mixed strategy z^* and the loss-ceiling $\tilde{\omega}^*$ for the player II can be obtained.

From the above discussion, we can summarize the process of the lexicographic method of matrix games with payoffs of triangular fuzzy numbers as follows.

Step 1: Construct the linear programming model according to Eq. (2.14), and solve it by using the simplex method of linear programming;
Step 2: Construct the linear programming model according to Eq. (2.23), and solve it by using the simplex method of linear programming;
Step 3: Construct the linear programming model according to Eq. (2.18), and solve it by using the simplex method of linear programming;
Step 4: Construct the linear programming model according to Eq. (2.25), and solve it by using the simplex method of linear programming;
Step 5: Obtain the solution of the matrix game $\tilde{\boldsymbol{A}}$ with payoffs of triangular fuzzy numbers, stop.

Example 2.3 Let us employ the above lexicographic method to solve the matrix game $\tilde{\boldsymbol{A}}_2$ with payoffs of triangular fuzzy numbers given in Example 2.2. Namely, the payoff matrix for the player I is given as follows:

$$\tilde{\boldsymbol{A}}_2 = \begin{matrix} & \beta_1 & \beta_2 \\ \delta_1 & (175, 180, 190) & (150, 156, 158) \\ \delta_2 & (80, 90, 100) & (175, 180, 190) \end{matrix}.$$

According to Eq. (2.14), the linear programming model can be constructed as follows:

$$\max\{\upsilon^m\}$$
$$\text{s.t.} \begin{cases} 175y_1 + 80y_2 \le \upsilon^l \\ 150y_1 + 175y_2 \le \upsilon^l \\ 180y_1 + 90y_2 \le \upsilon^m \\ 156y_1 + 180y_2 \le \upsilon^m \\ 190y_1 + 100y_2 \le \upsilon^r \\ 158y_1 + 190y_2 \le \upsilon^r \\ \upsilon^l \le \upsilon^m \le \upsilon^r \\ y_1 + y_2 = 1 \\ y_1 \ge 0, y_2 \ge 0 \\ \upsilon^l, \upsilon^m, \text{ and } \upsilon^r \text{ unrestricted in sign.} \end{cases}$$

Solving the above linear programming model by using the simplex method of linear programming, we can obtain its optimal solution $(\boldsymbol{y}^0, \upsilon^{l0}, \upsilon^{m*}, \upsilon^{r0})$ whose components are given as follows:

$$\boldsymbol{y}^0 = (0.789, 0.211)^{\mathrm{T}},\ \upsilon^{l0} = 61.408,\ \upsilon^{m*} = 161.06,\ \upsilon^{r0} = 163.073.$$

According to Eq. (2.23), the linear programming model can be constructed as follows:

$$\max\left\{\frac{\upsilon^l + \upsilon^r}{2}\right\}$$
$$\text{s.t.} \begin{cases} 175y_1 + 80y_2 \ge \upsilon^l \\ 150y_1 + 175y_2 \ge \upsilon^l \\ 180y_1 + 90y_2 \ge 161.06 \\ 156y_1 + 180y_2 \ge 161.06 \\ 190y_1 + 100y_2 \ge \upsilon^r \\ 158y_1 + 190y_2 \ge \upsilon^r \\ \upsilon^l \le 161.06 \le \upsilon^r \\ \upsilon^l \ge 61.408 \\ \upsilon^r \ge 163.073 \\ y_1 + y_2 = 1 \\ y_1 \ge 0, y_2 \ge 0 \\ \upsilon^l \text{ and } \upsilon^r \text{ unrestricted in sign.} \end{cases}$$

Solving the above linear programming model by using the simplex method of linear programming, we can obtain its optimal solution $(\boldsymbol{y}^*, \upsilon^{l*}, \upsilon^{r*})$ whose components are given as follows:

$$\boldsymbol{y}^* = (0.789, 0.211)^{\mathrm{T}},\ \upsilon^{l*} = 154.955,\ \upsilon^{r*} = 164.752.$$

Therefore, the maximin (or optimal) mixed strategy and the gain-floor for the player I are obtained as $\boldsymbol{y}^* = (0.789, 0.211)^{\mathrm{T}}$ and $\tilde{\upsilon}^* = (154.955, 161.06, 164.752)$, respectively.

Analogously, according to Eq. (2.18), the linear programming model can be obtained as follows:

$$\min\{\omega^m\}$$
$$\text{s.t.}\begin{cases} 175z_1 + 150z_2 \leq \omega^l \\ 80z_1 + 175z_2 \leq \omega^l \\ 180z_1 + 156z_2 \leq \omega^m \\ 90z_1 + 180z_2 \leq \omega^m \\ 190z_1 + 158z_2 \leq \omega^r \\ 100z_1 + 190z_2 \leq \omega^r \\ \omega^l \leq \omega^m \leq \omega^r \\ z_1 + z_2 = 1 \\ z_1 \geq 0, z_2 \geq 0 \\ \omega^l, \omega^m, \text{ and } \omega^r \text{ unrestricted in sign.} \end{cases}$$

Solving the above linear programming model by using the simplex method of linear programming, we can obtain its optimal solution $(\boldsymbol{z}^0, \omega^{l0}, \omega^{m*}, \omega^{r0})$ whose components are given as follows:

$$\boldsymbol{z}^0 = (0.211, 0.789)^{\mathrm{T}},\ \omega^{l0} = 158.908,\ \omega^{m*} = 161.06,\ \omega^{r0} = 339.62.$$

According to Eq. (2.25), the linear programming model can be obtained as follows:

$$\min\left\{\frac{\omega^l+\omega^r}{2}\right\}$$

$$\text{s.t.}\begin{cases} 175z_1+150z_2\leq\omega^l \\ 80z_1+175z_2\leq\omega^l \\ 180z_1+156z_2\leq 161.06 \\ 90z_1+180z_2\leq 161.06 \\ 190z_1+158z_2\leq\omega^r \\ 100z_1+190z_2\leq\omega^r \\ \omega^l\leq 161.06\leq\omega^r \\ \omega^l\leq 158.908 \\ \omega^r\leq 339.62 \\ z_1+z_2=1 \\ z_1\geq 0,\, z_2\geq 0 \\ \omega^l \text{ and } \omega^r \text{ unrestricted in sign.} \end{cases}$$

Solving the above linear programming model by using the simplex method of linear programming, we can obtain its optimal solution $(z^*, \omega^{l*}, \omega^{r*})$ whose components are given as follows:

$$z^*=(0.211, 0.789)^{\mathrm{T}},\ \omega^{l*}=155.275,\ \omega^{r*}=171.01.$$

Thus, the minimax (or optimal) mixed strategy and the loss-ceiling for the player II are obtained as $z^*=(0.211, 0.789)^{\mathrm{T}}$ and $\tilde{\omega}^*=(155.275, 161.06, 171.01)$, respectively.

Furthermore, the fuzzy equilibrium value of the matrix game $\tilde{\boldsymbol{A}}_2$ with payoffs of triangular fuzzy numbers can be obtained as follows:

$$\tilde{V}^*=\tilde{\upsilon}^*\wedge\tilde{\omega}^*=(155.275, 161.06, 164.752),$$

which means that the fuzzy value of the matrix game $\tilde{\boldsymbol{A}}_2$ with payoffs of triangular fuzzy numbers is "around 161.06". In other words, the player I's minimum reward is 155.275 while his/her maximum reward is 164.752. He/she could win any intermediate value x between 155.275 and 164.752 with the possibility $\mu_{\tilde{V}^*}(x)$ as follows:

$$\mu_{\tilde{V}^*}(x)=\begin{cases} \dfrac{x-155.275}{5.785} & \text{if } 155.275\leq x<161.06 \\ 1 & \text{if } x=161.06 \\ \dfrac{164.752-x}{3.692} & \text{if } 161.06<x\leq 164.752 \\ 0 & \text{else,} \end{cases}$$

depicted as in Fig. 2.7.

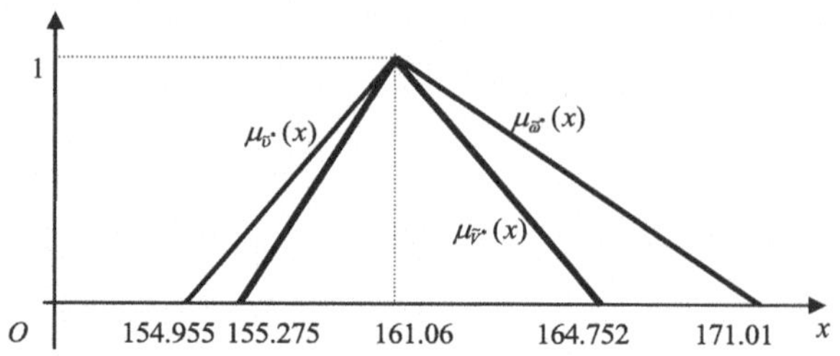

Fig. 2.7 The fuzzy equilibrium value $\tilde{V}^*$

2.6 Alfa-Cut-Based Primal-Dual Linear Programming Models of Matrix Games with Payoffs of Triangular Fuzzy Numbers

We firstly discuss a simple example of matrix games with payoffs of triangular fuzzy numbers.

Example 2.4 Let us consider a specific matrix game $\tilde{\boldsymbol{A}}^0$ with payoffs of triangular fuzzy numbers in which the player I's payoff matrix is given as follows:

$$\tilde{\boldsymbol{A}}^0 = \begin{matrix} \delta_1 \\ \delta_2 \\ \delta_3 \\ \delta_4 \end{matrix} \overset{\begin{matrix} \beta_1 & \qquad\beta_2 & \qquad\beta_3 & \qquad\beta_4 \end{matrix}}{\begin{pmatrix} (8,8.5,10) & \overline{(5,7,8)} & (14,16,18) & \overline{(5,7,8)} \\ (14,16,18) & \overline{(3.5,4,5)} & (-5,-3,-1) & \overline{(2,3,3.5)} \\ (11,12,14) & \overline{(5,7,8)} & (8,9,11) & \overline{(5,7,8)} \\ (-5,-3,-2) & \overline{(-1,0,2)} & (20,21,25) & \overline{(3.5,4,5)} \end{pmatrix}}.$$

By intuition observation or using the ranking relation of triangular fuzzy numbers and in the same way to crisp matrix games, it is easy to see from the minimax/maximin criteria [4, 26] that there are four pure strategy saddle points (δ_1, β_2), (δ_1, β_4), (δ_3, β_2), (δ_3, β_4) [or (1, 2), (1, 4), (3, 2), (3, 4)] and the matrix game $\tilde{\boldsymbol{A}}^0$ with payoffs of triangular fuzzy numbers has a fuzzy value $\tilde{V}^0 = (5, 7, 8)$, which is also a triangular fuzzy number. The fuzzy value means that the player I wins (5, 7, 8) whereas the player II loses (5, 7, 8) [or II wins $-\tilde{V}^0 = (-8, -7, -5)$] when I and II use the optimal pure strategies δ_1 (or δ_3) and β_2 (or β_4), respectively.

Unfortunately, in general, it is not always sure that there are pure strategy saddle points in matrix games with payoffs of triangular fuzzy numbers. Therefore, in the same way to crisp matrix games, we need to consider the players' mixed strategies $\boldsymbol{y}$ and $\boldsymbol{z}$ as stated in Sect. 1.2 or Sect. 2.3. Thus, stated as in Sect. 2.3.2, the player I's gain-floor $\tilde{\upsilon} = (\upsilon^l, \upsilon^m, \upsilon^r)$ and the player II's loss-ceiling $\tilde{\omega} = (\omega^l, \omega^m, \omega^r)$ are triangular fuzzy numbers. Moreover, it is always sure that $\tilde{\upsilon} \tilde{\leq} \tilde{\omega}$ according to Theorem 2.2.

In a similar way to Definition of the value of crisp matrix games [26], if $\tilde{\upsilon} = \tilde{\omega}$, then their common value is called the fuzzy value of the matrix game $\tilde{\boldsymbol{A}}$ with payoffs of triangular fuzzy numbers, i.e., $\tilde{V} = \tilde{\upsilon} = \tilde{\omega}$. In other words, the matrix game $\tilde{\boldsymbol{A}}$ with payoffs of triangular fuzzy numbers has a fuzzy value $\tilde{V}$. Obviously, $\tilde{V}$ is a triangular fuzzy number also, denoted by $\tilde{V} = (V^l, V^m, V^r)$.

2.6.1 *Interval-Valued Matrix Games Based on Alfa-Cut Sets of Triangular Fuzzy Numbers*

Stated as earlier, for any $\alpha \in [0, 1]$, α-cut sets of the triangular fuzzy numbers $\tilde{a}_{ij} = (a^l_{ij}, a^m_{ij}, a^r_{ij})$ are intervals, which are easily obtained by using Eq. (2.4) as follows:

$$\tilde{a}_{ij}(\alpha) = [a^L_{ij}(\alpha), a^R_{ij}(\alpha)] = [\alpha a^m_{ij} + (1-\alpha)a^l_{ij}, \alpha a^m_{ij} + (1-\alpha)a^r_{ij}]. \tag{2.26}$$

Let us consider an interval-valued matrix game $\tilde{\boldsymbol{A}}(\alpha)$ with the payoff matrix $\tilde{\boldsymbol{A}}(\alpha) = (\tilde{a}_{ij}(\alpha))_{m\times n}$, whose elements $\tilde{a}_{ij}(\alpha)$ are the intervals given by Eq. (2.26). $\tilde{a}_{ij}(\alpha)$ represents the interval-valued payoff of the player I when the players I and II use the pure strategies $\delta_i \in S_1$ and $\beta_j \in S_2$, respectively. Naturally, the player II's payoff is the interval $-\tilde{a}_{ij}(\alpha) = [-a^R_{ij}(\alpha), -a^L_{ij}(\alpha)]$ according to the arithmetic operations over intervals in Sect. 1.3.1.

Taking any value $a_{ij}(\alpha)$ in the interval-valued payoffs $\tilde{a}_{ij}(\alpha) = [a^L_{ij}(\alpha), a^R_{ij}(\alpha)]$, we consider a (crisp) matrix game $\boldsymbol{A}(\alpha)$ with the payoff matrix $\boldsymbol{A}(\alpha) = (a_{ij}(\alpha))_{m\times n}$. It is easy to from Eqs. (1.3) and (1.4) that the player I's gain-floor $v(\alpha)$ in the matrix game $\boldsymbol{A}(\alpha)$ is closely related to all $a_{ij}(\alpha)$. That is to say, $v(\alpha)$ is a function of $a_{ij}(\alpha)$ in the interval-valued payoffs $\tilde{a}_{ij}(\alpha)$, denoted by $v(\alpha) = \upsilon((a_{ij}(\alpha)))$. Similarly, the optimal mixed strategy $\boldsymbol{y}^*(\alpha)$ for the player I is a function of all $a_{ij}(\alpha)$ also, denoted by $\boldsymbol{y}^*(\alpha) = \boldsymbol{y}^*((a_{ij}(\alpha)))$.

In the same way to the above analysis, it is easy to see from Eqs. (1.6) and (1.7) that the loss-ceiling $\mu(\alpha)$ and corresponding optimal mixed strategy $\boldsymbol{z}^*(\alpha)$ for the player II in the matrix game $\boldsymbol{A}(\alpha)$ are functions of all $a_{ij}(\alpha)$ in the interval-valued payoffs $\tilde{a}_{ij}(\alpha)$, denoted by $\mu(\alpha) = \omega((a_{ij}(\alpha)))$ and $\boldsymbol{z}^*(\alpha) = \boldsymbol{z}^*((a_{ij}(\alpha)))$.

According to Eqs. (1.3) and (1.4), we can easily prove that the player I's gain-floor $\upsilon((a_{ij}(\alpha)))$ in the matrix game $\boldsymbol{A}(\alpha)$ is a non-decreasing function of all $a_{ij}(\alpha)$ in the interval-valued payoffs $\tilde{a}_{ij}(\alpha)$. In fact, for any $a_{ij}(\alpha)$ and $a'_{ij}(\alpha)$ in the interval-valued payoffs $\tilde{a}_{ij}(\alpha)$, if $a_{ij}(\alpha) \le a'_{ij}(\alpha)$, then

$$\sum_{i=1}^{m} y_i a_{ij}(\alpha) \le \sum_{i=1}^{m} y_i a'_{ij}(\alpha)$$

due to $y_i \ge 0$ $(i = 1, 2, \ldots, m)$ and $\sum_{i=1}^{m} y_i = 1$, where $\boldsymbol{y}$ is any mixed strategy of the player I. Hence, we have

$$\min_{1\leq j\leq n}\left\{\sum_{i=1}^{m}y_i a_{ij}(\alpha)\right\}\leq \min_{1\leq j\leq n}\left\{\sum_{i=1}^{m}y_i a'_{ij}(\alpha)\right\},$$

which directly infers that

$$\max_{y\in Y}\min_{1\leq j\leq n}\left\{\sum_{i=1}^{m}y_i a_{ij}(\alpha)\right\}\leq \max_{y\in Y}\min_{1\leq j\leq n}\left\{\sum_{i=1}^{m}y_i a'_{ij}(\alpha)\right\},$$

i.e.,

$$\upsilon((a_{ij}(\alpha)))\leq \upsilon((a'_{ij}(\alpha))),$$

where $\boldsymbol{A}'(\alpha)=(a'_{ij}(\alpha))_{m\times n}$ is the payoff matrix of the player I in the matrix game $\boldsymbol{A}'(\alpha)$.

According to the minimax theorem of matrix games [4, 26], the matrix game $\boldsymbol{A}(\alpha)$ has a value, denoted by $V(\alpha)=V((a_{ij}(\alpha)))$. Obviously, $V(\alpha)=v(\alpha)=\mu(\alpha)$. From the above discussion, $V((a_{ij}(\alpha)))$ is a non-decreasing function of all $a_{ij}(\alpha)$ in the interval-valued payoffs $\tilde{a}_{ij}(\alpha)$.

Stated as earlier, the value of the interval-valued matrix game $\tilde{\boldsymbol{A}}(\alpha)$ is an interval. The upper bound $v^R(\alpha)$ of the player I's gain-floor in the interval-valued matrix game $\tilde{\boldsymbol{A}}(\alpha)$ and corresponding optimal mixed strategy $\boldsymbol{y}^{R*}(\alpha)$ are $v^R(\alpha)=\upsilon^R((a^R_{ij}(\alpha)))$ and $\boldsymbol{y}^{R*}=\boldsymbol{y}^{R*}((a^R_{ij}(\alpha)))$, respectively. According to Eq. (1.5), $(v^R(\alpha),\boldsymbol{y}^{R*}(\alpha))$ is an optimal solution to the linear programming model as follows:

$$\max\{\upsilon^R(\alpha)\}$$
$$\text{s.t.}\begin{cases}\sum_{i=1}^{m}a^R_{ij}(\alpha)y^R_i(\alpha)\geq \upsilon^R(\alpha) & (j=1,2,\ldots,n)\\ \sum_{i=1}^{m}y^R_i(\alpha)=1\\ y^R_i(\alpha)\geq 0 & (i=1,2,\ldots,m)\\ \upsilon^R(\alpha)\ \text{unrestricted in sign,}\end{cases}\tag{2.27}$$

where $y^R_i(\alpha)$ $(i=1,2,\ldots,m)$ and $\upsilon^R(\alpha)$ are decision variables.

Without loss of generality [26], assume that $\upsilon^R(\alpha)>0$. Let

$$x^R_i(\alpha)=\frac{y^R_i(\alpha)}{\upsilon^R(\alpha)}\quad (i=1,2,\ldots,m).\tag{2.28}$$

Then, $x^R_i(\alpha)\geq 0$ $(i=1,2,\ldots,m)$ and

$$\sum_{i=1}^{m} x_i^R(\alpha) = \frac{1}{\upsilon^R(\alpha)}. \tag{2.29}$$

Combining with Eq. (2.26), Eq. (2.27) can be transformed into the linear programming model as follows:

$$\begin{aligned} &\min\left\{\sum_{i=1}^{m} x_i^R(\alpha)\right\} \\ &\text{s.t.}\begin{cases} \sum_{i=1}^{m} [\alpha a_{ij}^m + (1-\alpha)a_{ij}^r] x_i^R(\alpha) \geq 1 \quad (j = 1, 2, \ldots, n) \\ x_i^R(\alpha) \geq 0 \qquad (i = 1, 2, \ldots, m), \end{cases} \end{aligned} \tag{2.30}$$

where $x_i^R(\alpha)$ $(i = 1, 2, \ldots, m)$ are decision variables.

Solving Eq. (2.30) by using the simplex method of linear programming, we can obtain its optimal solution, denoted by $\boldsymbol{x}^{R*}(\alpha) = (x_1^{R*}(\alpha), x_2^{R*}(\alpha), \ldots, x_m^{R*}(\alpha))^{\mathrm{T}}$. According to Eqs. (2.28) and (2.29), the upper bound $v^R(\alpha)$ and the optimal mixed strategy $\boldsymbol{y}^{R*}(\alpha) = (y_1^{R*}(\alpha), y_2^{R*}(\alpha), \ldots, y_m^{R*}(\alpha))^{\mathrm{T}}$ can be obtained, respectively, where

$$v^R(\alpha) = \frac{1}{\sum_{i=1}^{m} x_i^{R*}(\alpha)} \tag{2.31}$$

and

$$y_i^{R*}(\alpha) = v^R(\alpha) x_i^{R*}(\alpha) \quad (i = 1, 2, \ldots, m). \tag{2.32}$$

Analogously, the lower bound $v^L(\alpha)$ of the player I's gain-floor in the interval-valued matrix game $\tilde{\boldsymbol{A}}(\alpha)$ and corresponding optimal mixed strategy $\boldsymbol{y}^{L*}(\alpha)$ are $v^L(\alpha) = \upsilon^L((a_{ij}^L(\alpha)))$ and $\boldsymbol{y}^{L*} = \boldsymbol{y}^{L*}((a_{ij}^L(\alpha)))$, respectively. Then, according to Eq. (1.5), $(v^L(\alpha), \boldsymbol{y}^{L*}(\alpha))$ is an optimal solution to the linear programming model as follows:

$$\begin{aligned} &\max\{\upsilon^L(\alpha)\} \\ &\text{s.t.}\begin{cases} \sum_{i=1}^{m} a_{ij}^L(\alpha) y_i^L(\alpha) \geq \upsilon^L(\alpha) \quad (j = 1, 2, \ldots, n) \\ \sum_{i=1}^{m} y_i^L(\alpha) = 1 \\ y_i^L(\alpha) \geq 0 \quad (i = 1, 2, \ldots, m) \\ \upsilon^L(\alpha) \text{ unrestricted in sign}, \end{cases} \end{aligned} \tag{2.33}$$

where $y_i^L(\alpha)$ $(i = 1, 2, \ldots, m)$ and $\upsilon^L(\alpha)$ are decision variables.

Without loss of generality [26], assume that $\upsilon^L(\alpha) > 0$. Let

$$x_i^L(\alpha) = \frac{y_i^L(\alpha)}{\upsilon^L(\alpha)} \quad (i = 1, 2, \ldots, m). \tag{2.34}$$

Then, $x_i^L(\alpha) \geq 0$ $(i = 1, 2, \ldots, m)$ and

$$\sum_{i=1}^{m} x_i^L(\alpha) = \frac{1}{\upsilon^L(\alpha)}. \tag{2.35}$$

Combining with Eq. (2.26), Eq. (2.33) can be transformed into the linear programming model as follows:

$$\begin{aligned} &\min\left\{\sum_{i=1}^{m} x_i^L(\alpha)\right\} \\ &\text{s.t.}\begin{cases} \sum_{i=1}^{m} [\alpha a_{ij}^m + (1-\alpha)a_{ij}^l] x_i^L(\alpha) \geq 1 & (j = 1, 2, \ldots, n) \\ x_i^L(\alpha) \geq 0 & (i = 1, 2, \ldots, m), \end{cases} \end{aligned} \tag{2.36}$$

where $x_i^L(\alpha)$ $(i = 1, 2, \ldots, m)$ are decision variables.

Solving Eq. (2.36) by using the simplex method of linear programming, we can obtain its optimal solution, denoted by $\boldsymbol{x}^{L*}(\alpha) = (x_1^{L*}(\alpha), x_2^{L*}(\alpha), \ldots, x_m^{L*}(\alpha))^{\mathrm{T}}$. According to Eqs. (2.34) and (2.35), the lower bound $v^L(\alpha)$ and the optimal mixed strategy $\boldsymbol{y}^{L*}(\alpha) = (y_1^{L*}(\alpha), y_2^{L*}(\alpha), \ldots, y_m^{L*}(\alpha))^{\mathrm{T}}$ can be obtained, respectively, where

$$v^L(\alpha) = \frac{1}{\sum_{i=1}^{m} x_i^{L*}(\alpha)} \tag{2.37}$$

and

$$y_i^{L*}(\alpha) = v^L(\alpha) x_i^{L*}(\alpha) \quad (i = 1, 2, \ldots, m). \tag{2.38}$$

Thus, the lower bound $v^L(\alpha)$ and upper bound $v^R(\alpha)$ and corresponding optimal mixed strategies can be obtained. Hence, the player I's gain-floor in the interval-valued matrix game $\tilde{\boldsymbol{A}}(\alpha)$ is obtained as an interval $\bar{v}(\alpha) = [v^L(\alpha), v^R(\alpha)]$, which is a α-cut set of $\tilde{\upsilon}$, i.e., $\bar{v}(\alpha) = \tilde{\upsilon}(\alpha)$.

In the same analysis, the upper bound $\mu^R(\alpha)$ of the player II's loss-ceiling in the interval-valued matrix game $\tilde{\boldsymbol{A}}(\alpha)$ and corresponding optimal mixed strategy $\boldsymbol{z}^{R*}(\alpha)$ are $\mu^R(\alpha) = \omega^R((a_{ij}^R(\alpha)))$ and $\boldsymbol{z}^{R*}(\alpha) = \boldsymbol{z}^{R*}((a_{ij}^R(\alpha)))$, respectively. According to Eq. (1.8), $(\mu^R(\alpha), \boldsymbol{z}^{R*}(\alpha))$ is an optimal solution to the linear programming model as follows:

$$\min\{\omega^R(\alpha)\}$$
$$\text{s.t.}\begin{cases} \sum_{j=1}^{n} a_{ij}^R(\alpha)z_j^R(\alpha) \le \omega^R(\alpha) \quad (i = 1, 2, \ldots, m) \\ \sum_{j=1}^{n} z_j^R(\alpha) = 1 \\ z_j^R(\alpha) \ge 0 \quad (j = 1, 2, \ldots, n) \\ \omega^R(\alpha) \text{ unrestricted in sign,} \end{cases} \tag{2.39}$$

where $\omega^R(\alpha)$ and $z_j^R(\alpha)$ $(j = 1, 2, \ldots, n)$ are decision variables.

Without loss of generality [26], assume that $\omega^R(\alpha) > 0$. Let

$$t_j^R(\alpha) = \frac{z_j^R(\alpha)}{\omega^R(\alpha)} \quad (j = 1, 2, \ldots, n), \tag{2.40}$$

thus, we have

$$\sum_{j=1}^{n} t_j^R(\alpha) = \frac{1}{\omega^R(\alpha)}. \tag{2.41}$$

Combining with Eq. (2.26), Eq. (2.39) can be converted into the linear programming model as follows:

$$\max\left\{\sum_{j=1}^{n} t_j^R(\alpha)\right\}$$
$$\text{s.t.}\begin{cases} \sum_{j=1}^{n} [\alpha a_{ij}^m + (1-\alpha)a_{ij}^r] t_j^R(\alpha) \le 1 \quad (i = 1, 2, \ldots, m) \\ t_j^R(\alpha) \ge 0 \qquad (j = 1, 2, \ldots, n), \end{cases} \tag{2.42}$$

where $t_j^R(\alpha)$ $(j = 1, 2, \ldots, n)$ are decision variables.

Solving Eq. (2.42) by using the simplex method of linear programming, we can obtain its optimal solution, denoted by $\boldsymbol{t}^{R*}(\alpha) = (t_1^{R*}(\alpha), t_2^{R*}(\alpha), \ldots, t_n^{R*}(\alpha))^{\mathrm{T}}$. According to Eqs. (2.40) and (2.41), the upper bound $\mu^R(\alpha)$ and the optimal mixed strategy $\boldsymbol{z}^{R*}(\alpha) = (z_1^{R*}(\alpha), z_2^{R*}(\alpha), \ldots, z_n^{R*}(\alpha))^{\mathrm{T}}$ can be obtained, respectively, where

$$\mu^R(\alpha) = \frac{1}{\sum_{j=1}^{n} t_j^{R*}(\alpha)} \tag{2.43}$$

and

$$z_j^{R*}(\alpha) = \mu^R(\alpha) t_j^{R*}(\alpha) \quad (j = 1, 2, \ldots, n). \tag{2.44}$$

Analogously, the lower bound $\mu^L(\alpha)$ of the player II's loss-ceiling in the interval-valued matrix game $\tilde{\boldsymbol{A}}(\alpha)$ and corresponding optimal mixed strategy $\boldsymbol{z}^{L*}(\alpha)$ are $\mu^L(\alpha) = \omega^L((a_{ij}^L(\alpha)))$ and $\boldsymbol{z}^{L*}(\alpha) = \boldsymbol{z}^{L*}((a_{ij}^L(\alpha)))$, respectively. Then, according to Eq. (1.8), $(\mu^L(\alpha), \boldsymbol{z}^{L*}(\alpha))$ is an optimal solution to the linear programming model as follows:

$$\min\{\omega^L(\alpha)\}$$
$$\text{s.t.}\begin{cases}\sum_{j=1}^{n} a_{ij}^L(\alpha) z_j^L(\alpha) \le \omega^L(\alpha) \quad (i = 1, 2, \ldots, m) \\ \sum_{j=1}^{n} z_j^L(\alpha) = 1 \\ z_j^L(\alpha) \ge 0 \quad (j = 1, 2, \ldots, n) \\ \omega^L(\alpha) \text{ unrestricted in sign,}\end{cases} \tag{2.45}$$

where $\omega^L(\alpha)$ and $z_j^L(\alpha)$ $(j = 1, 2, \ldots, n)$ are decision variables.

Without loss of generality [26], assume that $\omega^L(\alpha) > 0$. Let

$$t_j^L(\alpha) = \frac{z_j^L(\alpha)}{\omega^L(\alpha)} \quad (j = 1, 2, \ldots, n), \tag{2.46}$$

then

$$\sum_{j=1}^{n} t_j^L(\alpha) = \frac{1}{\omega^L(\alpha)}. \tag{2.47}$$

Combining with Eq. (2.26), Eq. (2.45) can be converted into the linear programming model as follows:

$$\max\left\{\sum_{j=1}^{n} t_j^L(\alpha)\right\}$$
$$\text{s.t.}\begin{cases}\sum_{j=1}^{n} [\alpha a_{ij}^m + (1-\alpha) a_{ij}^l] t_j^L(\alpha) \le 1 \quad (i = 1, 2, \ldots, m) \\ t_j^L(\alpha) \ge 0 \qquad (j = 1, 2, \ldots, n),\end{cases} \tag{2.48}$$

where $t_j^L(\alpha)$ $(j = 1, 2, \ldots, n)$ are decision variables.

Solving Eq. (2.48) by using the simplex method of linear programming, we can obtain its optimal solution, denoted by $\boldsymbol{t}^{L*}(\alpha) = (t_1^{L*}(\alpha), t_2^{L*}(\alpha), \ldots, t_n^{L*}(\alpha))^{\mathrm{T}}$. According to Eqs. (2.46) and (2.47), the lower bound $\mu^L(\alpha)$ and the optimal mixed strategy $\boldsymbol{z}^{L*}(\alpha) = (z_1^{L*}(\alpha), z_2^{L*}(\alpha), \ldots, z_n^{L*}(\alpha))^{\mathrm{T}}$ can be obtained, respectively, where

$$\mu^L(\alpha) = \frac{1}{\sum_{j=1}^n t_j^{L*}(\alpha)} \tag{2.49}$$

and

$$z_j^{L*}(\alpha) = \mu^L(\alpha) t_j^{L*}(\alpha) \quad (j = 1, 2, \ldots, n). \tag{2.50}$$

Thus, the lower bound $\mu^L(\alpha)$ and upper bound $\mu^R(\alpha)$ and corresponding optimal mixed strategies for the player II can be obtained. Hereby, the player II's loss-ceiling in the interval-valued matrix game $\tilde{\boldsymbol{A}}(\alpha)$ is obtained as an interval $\bar{\mu}(\alpha) = [\mu^L(\alpha), \mu^R(\alpha)]$, which is a α-cut set of $\tilde{\omega}$, i.e., $\bar{\mu}(\alpha) = \tilde{\omega}(\alpha)$.

It is easy to see that Eqs. (2.30) and (2.42) are a pair of primal-dual linear programming models. Therefore, the minimum of $\sum_{i=1}^m x_i^R(\alpha)$ (i.e., the maximum of $\upsilon^R(\alpha)$) is equal to the maximum of $\sum_{j=1}^n t_j^R(\alpha)$ (i.e., the minimum of $\omega^R(\alpha)$) by the duality theorem of linear programming, i.e.,

$$v^R(\alpha) = \mu^R(\alpha).$$

In the same way, Eqs. (2.36) and (2.48) are a pair of primal-dual linear programming models. Hence, we have

$$v^L(\alpha) = \mu^L(\alpha).$$

Therefore, the player I's gain-floor $\bar{v}(\alpha) = [v^L(\alpha), v^R(\alpha)]$ is equal to the player II's loss-ceiling $\bar{\mu}(\alpha) = [\mu^L(\alpha), \mu^R(\alpha)]$, i.e., $\bar{v}(\alpha) = \bar{\mu}(\alpha)$. Namely, the players' gain-floor and loss-ceiling have a common interval-type value. According to Definition of the value of matrix games [26], the interval-valued matrix game $\tilde{\boldsymbol{A}}(\alpha)$ has an interval-type value, denoted by the interval $\bar{V}(\alpha) = [V^L(\alpha), V^R(\alpha)]$, where $\bar{V}(\alpha) = \bar{v}(\alpha) = \bar{\mu}(\alpha)$. Essentially, $\bar{V}(\alpha)$ is a α-cut set of $\tilde{V}$ of the matrix game $\tilde{\boldsymbol{A}}$ with payoffs of triangular fuzzy numbers. Noticing the fact that $\bar{V}(\alpha) = \tilde{\upsilon}(\alpha) = \tilde{\omega}(\alpha)$ for any $\alpha \in [0, 1]$. According to the concept of α-cuts and the representation theorem for fuzzy sets [5], we directly have $\tilde{V} = \tilde{\upsilon} = \tilde{\omega}$, which infers that the player I's gain-floor $\tilde{\upsilon}$ is equal to the player II's loss-ceiling $\tilde{\omega}$ (or the players' gain-floor and loss-ceiling have a common value) and hereby the matrix game $\tilde{\boldsymbol{A}}$ with payoffs of triangular fuzzy numbers has the fuzzy value $\tilde{V}$, which is also a triangular fuzzy number as stated in Sect. 2.2.

Example 2.5 Let us again consider the matrix game $\tilde{\boldsymbol{A}}^0$ with payoffs of triangular fuzzy numbers, which is given in Example 2.4.

For any $\alpha \in [0, 1]$, we can obtain the interval-valued matrix game $\tilde{\boldsymbol{A}}^0(\alpha)$ whose interval-valued payoff matrix is given as follows:

$$\tilde{A}^0(\alpha) = \begin{matrix} \\ \delta_1 \\ \delta_2 \\ \delta_3 \\ \delta_4 \end{matrix} \begin{matrix} \beta_1 & \beta_2 & \beta_3 & \beta_4 \\ \left(\begin{matrix} [8+0.5\alpha, 10-1.5\alpha] \\ [14+2\alpha, 18-2\alpha] \\ [11+\alpha, 14-2\alpha] \\ [-5+2\alpha, -1-2\alpha] \end{matrix}\right. & \begin{matrix} \underline{[5+2\alpha, 8-\alpha]} \\ [3.5+0.5\alpha, 5-\alpha] \\ \underline{[5+2\alpha, 8-\alpha]} \\ [-1+\alpha, 2-\alpha] \end{matrix} & \begin{matrix} [14+2\alpha, 18-2\alpha] \\ [-5+2\alpha, -1-2\alpha] \\ [8+\alpha, 11-2\alpha] \\ [20+\alpha, 25-4\alpha] \end{matrix} & \left.\begin{matrix} \underline{[5+2\alpha, 8-\alpha]} \\ [2+\alpha, 3.5-0.5\alpha] \\ \underline{[5+2\alpha, 8-\alpha]} \\ [3.5+0.5\alpha, 5-\alpha] \end{matrix}\right) \end{matrix}.$$

According to the minimax/maximin criteria and the ranking methods of intervals, it is easy to see that the players' gain-floor and loss-ceiling have a common interval-type value, i.e., $\bar{v}^0(\alpha) = \bar{\rho}^0(\alpha) = [5+2\alpha, 8-\alpha]$. Therefore, there are still the four pure strategy saddle points (δ_1, β_2), (δ_1, β_4), (δ_3, β_2), (δ_3, β_4) [or (1, 2), (1, 4), (3, 2), (3, 4)] and the interval-valued matrix game $\tilde{A}^0(\alpha)$ has an interval-type value $\bar{V}^0(\alpha) = [5+2\alpha, 8-\alpha]$. Noticing that $\alpha \in [0, 1]$ is arbitrary. Hence, the player I's gain-floor in the aforementioned matrix game $\tilde{A}^0$ with payoffs of triangular fuzzy numbers is equal to the player II's loss-ceiling, i.e., $\tilde{\upsilon}^0 = \tilde{\omega}^0 = (5, 7, 8)$. Thus, the matrix game $\tilde{A}^0$ with payoffs of triangular fuzzy numbers has a fuzzy value $\tilde{V}^0$ at the pure strategy saddle points (δ_1, β_2), (δ_1, β_4), (δ_3, β_2), and (δ_3, β_4), where $\tilde{V}^0 = \tilde{\upsilon}^0 = \tilde{\omega}^0 = (5, 7, 8)$. Obviously, these results are the same as those obtained in Example 2.4.

Likewise, for the aforementioned interval-valued matrix game $\tilde{A}^0(\alpha)$, according to Eqs. (2.30), (2.36), (2.42), and (2.48), we can easily obtain the player I's gain-floor $\bar{v}^0(\alpha) = [5+2\alpha, 8-\alpha]$ and optimal mixed strategy $\boldsymbol{y}^* = (0.5, 0, 0.5, 0)^{\mathrm{T}}$ as well as the player II's loss-ceiling $\bar{\mu}^0(\alpha) = [5+2\alpha, 8-\alpha]$ and optimal mixed strategy $\boldsymbol{z}^* = (0, 0.5, 0, 0.5)^{\mathrm{T}}$. Then, the interval-valued matrix game $\tilde{A}^0(\alpha)$ has an interval-type value $\bar{V}^0(\alpha)$, where $\bar{V}^0(\alpha) = \bar{v}^0(\alpha) = \bar{\mu}^0(\alpha)$. Hereby, the matrix game $\tilde{A}^0$ with payoffs of triangular fuzzy numbers has the fuzzy value $\tilde{V}^0 = (5, 7, 8)$ and corresponding optimal mixed strategies for the players I and II are $\boldsymbol{y}^* = (0.5, 0, 0.5, 0)^{\mathrm{T}}$ and $\boldsymbol{z}^* = (0, 0.5, 0, 0.5)^{\mathrm{T}}$, respectively, where $\tilde{V}^0 = \tilde{\upsilon}^0 = \tilde{\omega}^0 = (5, 7, 8)$.

Example 2.6 Let us use the proposed method in this section to solve the specific matrix game $\tilde{A}_2$ with payoffs of triangular fuzzy numbers given in Example 2.2. The payoff matrix of the player I is $\tilde{A}_2$ given as in Example 2.2 and the pure and mixed strategies of the players I and II are crisp.

According to Eqs. (2.30) and (2.42), the linear programming models are constructed as follows:

$$\min\{x_1^R(\alpha) + x_2^R(\alpha)\}$$
$$\text{s.t.} \begin{cases} [180\alpha + 190(1-\alpha)]x_1^R(\alpha) + [90\alpha + 100(1-\alpha)]x_2^R(\alpha) \geq 1 \\ [156\alpha + 158(1-\alpha)]x_1^R(\alpha) + [180\alpha + 190(1-\alpha)]x_2^R(\alpha) \geq 1 \\ x_1^R(\alpha) \geq 0, x_2^R(\alpha) \geq 0 \end{cases}$$

Table 2.1 Upper and lower bounds of interval-type values of the interval-valued matrix games and the players' optimal strategies

α	0	0.1	0.2	0.3
$V^R(\alpha)$	166.3934	165.8317	165.2757	164.7257
$(y_1^{R*}(\alpha), y_2^{R*}(\alpha))^T$	(0.7377, 0.2623)	(0.7426, 0.2574)	(0.7475, 0.2525)	(0.7525, 0.2475)
$(z_1^{R*}(\alpha), z_2^{R*}(\alpha))^T$	(0.2623, 0.7377)	(0.2574, 0.7426)	(0.2525, 0.7475)	(0.2475, 0.7525)
$V^L(\alpha)$	155.2083	155.7927	156.3771	156.9615
$(y_1^{L*}(\alpha), y_2^{L*}(\alpha))^T$	(0.7917, 0.2083)	(0.7915, 0.2085)	(0.7912, 0.2088)	(0.7910, 0.2090)
$(z_1^{L*}(\alpha), z_2^{L*}(\alpha))^T$	(0.2083, 0.7917)	(0.2085, 0.7915)	(0.2088, 0.7912)	(0.2090, 0.7910)
$\tilde{V}(\alpha) = [V^L(\alpha), V^R(\alpha)]$	[155.2083, 166.3934]	[155.7927, 165.8317]	[156.3771, 165.2757]	[156.9615, 164.7257]
α	0.4	0.5	0.6	0.7
$V^R(\alpha)$	164.1818	163.6441	163.1126	162.5876
$(y_1^{R*}(\alpha), y_2^{R*}(\alpha))^T$	(0.7576, 0.2424)	(0.7627, 0.2373)	(0.7679, 0.2321)	(0.7732, 0.2268)
$(z_1^{R*}(\alpha), z_2^{R*}(\alpha))^T$	(0.2424, 0.7576)	(0.2373, 0.7627)	(0.2321, 0.7679)	(0.2268, 0.7732)
$V^L(\alpha)$	157.5459	158.1303	158.7148	159.2992
$(y_1^{L*}(\alpha), y_2^{L*}(\alpha))^T$	(0.7908, 0.2092)	(0.7906, 0.2094)	(0.7904, 0.2096)	(0.7902, 0.2098)
$(z_1^{L*}(\alpha), z_2^{L*}(\alpha))^T$	(0.2092, 0.7908)	(0.2094, 0.7906)	(0.2096, 0.7904)	(0.2098, 0.7902)
$\tilde{V}(\alpha) = [V^L(\alpha), V^R(\alpha)]$	[157.5459, 164.1818]	[158.1303, 163.6441]	[158.7148, 163.1126]	[159.2992, 162.5876]
α	0.8	0.9	1.0	
$V^R(\alpha)$	162.0692	161.5575	161.0526	
$(y_1^{R*}(\alpha), y_2^{R*}(\alpha))^T$	(0.7785, 0.2215)	(0.7840, 0.2160)	(0.7895, 0.2105)	
$(z_1^{R*}(\alpha), z_2^{R*}(\alpha))^T$	(0.2215, 0.7785)	(0.2160, 0.7840)	(0.2105, 0.7895)	
$V^L(\alpha)$	159.8837	160.4682	161.0526	
$(y_1^{L*}(\alpha), y_2^{L*}(\alpha))^T$	(0.7899, 0.2101)	(0.7897, 0.2103)	(0.7895, 0.2105)	
$(z_1^{L*}(\alpha), z_2^{L*}(\alpha))^T$	(0.2101, 0.7899)	(0.2103, 0.7897)	(0.2105, 0.7895)	
$\tilde{V}(\alpha) = [V^L(\alpha), V^R(\alpha)]$	[159.8837, 162.0692]	[160.4682, 161.5575]	161.0526	

and

$$\max\{t_1^R(\alpha)+t_2^R(\alpha)\}$$
$$\text{s.t.}\begin{cases}[180\alpha+190(1-\alpha)]t_1^R(\alpha)+[156\alpha+158(1-\alpha)]t_2^R(\alpha)\leq 1\\ [90\alpha+100(1-\alpha)]t_1^R(\alpha)+[180\alpha+190(1-\alpha)]t_2^R(\alpha)\leq 1\\ t_1^R(\alpha)\geq 0, t_2^R(\alpha)\geq 0,\end{cases}$$

where $x_1^R(\alpha)$, $x_2^R(\alpha)$, $t_1^R(\alpha)$, and $t_2^R(\alpha)$ are decision variables.

For some given special values of $\alpha\in[0,1]$, solving the above two linear programming models by using the simplex method of linear programming, we can obtain their optimal solutions $\boldsymbol{x}^{R*}(\alpha)=(x_1^{R*}(\alpha),x_2^{R*}(\alpha))^{\mathrm{T}}$ and $\boldsymbol{t}^{R*}(\alpha)=(t_1^{R*}(\alpha), t_2^{R*}(\alpha))^{\mathrm{T}}$, respectively. Combining with Eqs. (2.31), (2.32), (2.43), and (2.44), we obtain the upper bounds of the interval-type values of the interval-valued matrix games and corresponding optimal strategies for the players I and II, depicted as in Table 2.1.

Analogously, according to (2.36) and (2.48), the linear programming models are constructed as follows:

$$\min\{x_1^L(\alpha)+x_2^L(\alpha)\}$$
$$\text{s.t.}\begin{cases}[180\alpha+175(1-\alpha)]x_1^L(\alpha)+[90\alpha+80(1-\alpha)]x_2^L(\alpha)\geq 1\\ [156\alpha+150(1-\alpha)]x_1^L(\alpha)+[180\alpha+175(1-\alpha)]x_2^L(\alpha)\geq 1\\ x_1^L(\alpha)\geq 0, x_2^L(\alpha)\geq 0\end{cases}$$

and

$$\max\{t_1^L(\alpha)+t_2^L(\alpha)\}$$
$$\text{s.t.}\begin{cases}[180\alpha+175(1-\alpha)]t_1^L(\alpha)+[156\alpha+150(1-\alpha)]t_2^L(\alpha)\leq 1\\ [90\alpha+80(1-\alpha)]t_1^L(\alpha)+[180\alpha+175(1-\alpha)]t_2^L(\alpha)\leq 1\\ t_1^L(\alpha)\geq 0, t_2^L(\alpha)\geq 0,\end{cases}$$

where $x_1^L(\alpha)$, $x_2^L(\alpha)$, $t_1^L(\alpha)$, and $t_2^L(\alpha)$ are decision variables.

For the given special values of $\alpha\in[0,1]$, solving the above linear programming models by using the simplex method of linear programming, we can obtain their optimal solutions $\boldsymbol{x}^{L*}(\alpha)=(x_1^{L*}(\alpha),x_2^{L*}(\alpha))^{\mathrm{T}}$ and $\boldsymbol{t}^{L*}(\alpha)=(t_1^{L*}(\alpha),t_2^{L*}(\alpha))^{T}$, respectively. Combining with Eqs. (2.37), (2.38), (2.49), and (2.50), we obtain the lower bounds of the interval-type values of the interval-valued matrix games and corresponding optimal strategies for the players I and II, depicted as in Table 2.1.

For $\alpha=1$, it is easy to see from Table 2.1 that the value of the interval-valued matrix game is $\tilde{V}(1)=161.0526$ when the player I employs the optimal strategy $(0.7895, 0.2105)^{\mathrm{T}}$ and the player II employs the optimal strategy $(0.2105, 0.7895)^{\mathrm{T}}$, respectively. It is noticed that the upper and lower bounds of the interval-type value of the interval-valued matrix game are identical, i.e., $V^L(1) = V^R(1) = 161.0526$. Namely, the interval-type value $\tilde{V}(1)$ degenerates to the real number 161.0526. Moreover, the player I's optimal strategies $\boldsymbol{y}^{R*}(1) = (y_1^{R*}(1), y_2^{R*}(1))^{\mathrm{T}}$ and $\boldsymbol{y}^{L*}(1) = (y_1^{L*}(1), y_2^{L*}(1))^{\mathrm{T}}$ are identical, i.e., $\boldsymbol{y}^{R*}(1) = \boldsymbol{y}^{L*}(1) = (0.7895, 0.2105)^{\mathrm{T}}$. The player II's optimal strategies $\boldsymbol{z}^{R*}(1) = (z_1^{R*}(1), z_2^{R*}(1))^{\mathrm{T}}$ and $\boldsymbol{z}^{L*}(1) = (z_1^{L*}(1), z_2^{L*}(1))^{\mathrm{T}}$ are identical, i.e., $\boldsymbol{z}^{R*}(1) = \boldsymbol{z}^{L*}(1) = (0.2105, 0.7895)^{\mathrm{T}}$.

In the same way, for $\alpha = 0$, it is easy to see from Table 2.1 that the value of the interval-valued matrix game is the interval $\tilde{V}(0) = [155.2083, 166.3934]$. The player I wins (i.e., the player II loses) the upper bound $V^R(0) = 166.3934$ of the value $\tilde{V}(0)$ when the player I employs the optimal strategy $\boldsymbol{y}^{R*}(0) = (0.7377, 0.2623)^{\mathrm{T}}$ and the player II employs the optimal strategy $\boldsymbol{z}^{R*}(0) = (0.2623, 0.7377)^{\mathrm{T}}$, respectively. The player I wins (i.e., the player II loses) the lower bound $V^L(0) = 155.2083$ of the value $\tilde{V}(0)$ when the player I employs the optimal strategy $\boldsymbol{y}^{L*}(0) = (0.7917, 0.2083)^{\mathrm{T}}$ and the player II employs the optimal strategy $\boldsymbol{z}^{L*}(0) = (0.2083, 0.7917)^{\mathrm{T}}$, respectively.

For $\alpha = 0.6$, it is easy to see from Table 2.1 that the value of the interval-valued matrix game is the interval $\tilde{V}(0.6) = [158.7158, 163.1126]$. The player I wins (i.e., the player II loses) the upper bound $V^R(0.6) = 163.1126$ of the value $\tilde{V}(0.6)$ when the player II employs the optimal strategy $\boldsymbol{y}^{R*}(0.6) = (0.7679, 0.2321)^{\mathrm{T}}$ and the player II employs the optimal strategy $\boldsymbol{z}^{R*}(0.6) = (0.2321, 0.7679)^{\mathrm{T}}$, respectively. Likewise, the player I wins (i.e., the player II loses) the lower bound $V^L(0.6) = 158.7148$ of the value $\tilde{V}(0.6)$ when the player I employs the optimal strategy $\boldsymbol{y}^{L*}(0.6) = (0.7904, 0.2096)^{\mathrm{T}}$ and the player II employs the optimal strategy $\boldsymbol{z}^{L*}(0.6) = (0.2096, 0.7904)^{\mathrm{T}}$, respectively. The obtained results in Table 2.1 for the other values $\alpha \in [0, 1]$ are similarly explained.

2.6.2 *Linear Programming Method of Matrix Games with Payoffs of Triangular Fuzzy Numbers*

Usually, computing fuzzy values of matrix games with payoffs of triangular fuzzy numbers is not easier than that in Example 2.5. In the sequent, we focus on developing an effective and a simple method which can explicitly and quickly compute fuzzy values of matrix games $\tilde{\boldsymbol{A}}$ with payoffs of triangular fuzzy numbers.

For $\alpha = 1$, according to Eqs. (2.30) and (2.48), the linear programming models are constructed as follows:

$$\min\left\{\sum_{i=1}^{m} x_i^R(1)\right\}$$
$$\text{s.t.}\begin{cases}\sum_{i=1}^{m} a_{ij}^m x_i^R(1) \geq 1 & (j = 1, 2, \ldots, n)\\ x_i^R(1) \geq 0 & (i = 1, 2, \ldots, m)\end{cases} \tag{2.51}$$

and

$$\max\left\{\sum_{j=1}^{n} t_j^L(1)\right\}$$
$$\text{s.t.}\begin{cases}\sum_{j=1}^{n} a_{ij}^m t_j^L(1) \leq 1 & (i = 1, 2, \ldots, m)\\ t_j^L(1) \geq 0 & (j = 1, 2, \ldots, n),\end{cases} \tag{2.52}$$

where $x_i^R(1)$ and $t_j^L(1)$ $(i = 1, 2, \ldots, m; j = 1, 2, \ldots, n)$ are decision variables.

Obviously, Eqs. (2.51) and (2.52) are a pair of primal-dual linear programming models. Then, the minimum of $\sum_{i=1}^{m} x_i^R(1)$ (i.e., the maximum of $\upsilon^R(1)$) is equal to the maximum of $\sum_{j=1}^{n} t_j^L(1)$ (i.e., the minimum of $\omega^L(1)$) by the duality theorem of linear programming [26], i.e., $v^R(1) = \mu^L(1)$.

Analogously, for $\alpha = 1$, according to Eqs. (2.36) and (2.42), the linear programming models are constructed as follows:

$$\min\left\{\sum_{i=1}^{m} x_i^L(1)\right\}$$
$$\text{s.t.}\begin{cases}\sum_{i=1}^{m} a_{ij}^m x_i^L(1) \geq 1 & (j = 1, 2, \ldots, n)\\ x_i^L(1) \geq 0 & (i = 1, 2, \ldots, m)\end{cases} \tag{2.53}$$

and

$$\max\left\{\sum_{j=1}^{n} t_j^R(1)\right\}$$
$$\text{s.t.}\begin{cases}\sum_{j=1}^{n} a_{ij}^m t_j^R(1) \leq 1 & (i = 1, 2, \ldots, m)\\ t_j^R(1) \geq 0 & (j = 1, 2, \ldots, n),\end{cases} \tag{2.54}$$

where $x_i^L(1)$ and $t_j^R(1)$ $(i = 1, 2, \ldots, m; j = 1, 2, \ldots, n)$ are decision variables.

Obviously, Eqs. (2.53) and (2.54) are a pair of primal-dual linear programming models. According to the duality theorem of linear programming, we have

$\nu^L(1) = \mu^R(1)$. Combining with the above discussion, it directly follows that $\nu^L(1) = \nu^R(1) = \mu^L(1) = \mu^R(1)$. Thus, $[\nu^L(1), \nu^R(1)] = [\mu^L(1), \mu^R(1)]$ degenerates to a real number. Hence, $V^L(1) = V^R(1) = \nu^L(1)$, i.e., $\tilde{V}(1)$ is a real number. It is derived from the notation of the triangular fuzzy number $\tilde{V} = (V^l, V^m, V^r)$ that $V^m = V^L(1) = V^R(1)$. Namely, the mean of the fuzzy value $\tilde{V}$ can be directly obtained by solving one of Eqs. (2.51)–(2.54).

In the same way, for $\alpha = 0$, according to Eqs. (2.30) and (2.42), the linear programming models are constructed as follows:

$$\begin{aligned} &\min\left\{\sum_{i=1}^{m} x_i^R(0)\right\} \\ &\text{s.t.}\begin{cases} \sum_{i=1}^{m} a_{ij}^r x_i^R(0) \geq 1 & (j = 1, 2, \ldots, n) \\ x_i^R(0) \geq 0 & (i = 1, 2, \ldots, m) \end{cases} \end{aligned} \tag{2.55}$$

and

$$\begin{aligned} &\max\left\{\sum_{j=1}^{n} t_j^R(0)\right\} \\ &s.t.\begin{cases} \sum_{j=1}^{n} a_{ij}^r t_j^R(0) \leq 1 & (i = 1, 2, \ldots, m) \\ t_j^R(0) \geq 0 & (j = 1, 2, \ldots, n), \end{cases} \end{aligned} \tag{2.56}$$

which infer that $\nu^R(0) = \mu^R(0)$.

Analogously, according to Eqs. (2.36) and (2.48), the linear programming models are constructed as follows:

$$\begin{aligned} &\min\left\{\sum_{i=1}^{m} x_i^L(0)\right\} \\ &\text{s.t.}\begin{cases} \sum_{i=1}^{m} a_{ij}^l x_i^L(0) \geq 1 & (j = 1, 2, \ldots, n) \\ x_i^L(0) \geq 0 & (i = 1, 2, \ldots, m) \end{cases} \end{aligned} \tag{2.57}$$

and

$$\begin{aligned} &\max\left\{\sum_{j=1}^{n} t_j^L(0)\right\} \\ &\text{s.t.}\begin{cases} \sum_{j=1}^{n} a_{ij}^l t_j^L(0) \leq 1 & (i = 1, 2, \ldots, m) \\ t_j^L(0) \geq 0 & (j = 1, 2, \ldots, n), \end{cases} \end{aligned} \tag{2.58}$$

which infer that $v^L(0) = \mu^L(0)$.

It is easily derived from the above discussion that $\tilde{V}(0) = \tilde{v}(0) = \tilde{\mu}(0)$. According to the notation of the triangular fuzzy number $\tilde{V} = (V^l, V^m, V^r)$, it follows that $V^l = V^L(0) = v^L(0)$ and $V^r = V^R(0) = v^R(0)$, which mean that the lower and upper bounds of the fuzzy value $\tilde{V}$ can be directly obtained by solving either Eqs. (2.55) and (2.57) or Eqs. (2.56) and (2.58). Obviously, $\tilde{V}(0) = [V^L(0), V^R(0)] = [V^l, V^r]$.

Thus, according to Eq. (2.4), any α-cut set of the fuzzy value $\tilde{V}$ of the matrix game $\tilde{\boldsymbol{A}}$ with payoffs of triangular fuzzy numbers can be obtained as

$$[V^L(\alpha), V^R(\alpha)] = [\alpha V^m + (1-\alpha)V^l, \alpha V^m + (1-\alpha)V^r].$$

Hereby, according to Eq. (2.5) or the representation theorem for the fuzzy set [5], the fuzzy value $\tilde{V}$ can be expressed as

$$\tilde{V} = \bigcup_{\alpha\in[0,1]} \{\alpha \otimes \tilde{V}(\alpha)\} = \bigcup_{\alpha\in[0,1]} \{\alpha \otimes [\alpha V^m + (1-\alpha)V^l, \alpha V^m + (1-\alpha)V^r]\},$$

which means that $\tilde{V}$ can be explicitly obtained by using both its 1-cut set and 0-cut set of fuzzy payoffs.

2.6.3 Computational Analysis of a Real Example

Let us continue to consider the specific matrix game $\tilde{\boldsymbol{A}}_2$ with payoffs of triangular fuzzy numbers given in Example 2.2. The players' pure and mixed strategies are crisp and the player I' payoff matrix is $\tilde{\boldsymbol{A}}_2$ as stated in Example 2.2.

1. Computational results obtained by the proposed Alfa-cut-based primal-dual linear programming method

   Using Eq. (2.51), the linear programming model is constructed as follows:

$$\min\{x_1^R(1) + x_2^R(1)\}$$
$$\text{s.t.}\begin{cases} 180x_1^R(1) + 90x_2^R(1) \geq 1 \\ 156x_1^R(1) + 180x_2^R(1) \geq 1 \\ x_1^R(1) \geq 0,\ x_2^R(1) \geq 0, \end{cases}$$

where $x_1^R(1)$ and $x_2^R(1)$ are decision variables. Solving the above linear programming model by using the simplex method of linear programming, we obtain its optimal solution $\boldsymbol{x}^{R*}(1) = (x_1^{R*}(1), x_2^{R*}(1))^{\mathrm{T}}$, where

$$x_1^{R*}(1) = \frac{1}{204} \approx 0.0049, \quad x_2^{R*}(1) = \frac{1}{765} \approx 0.0013.$$

According to Eqs. (2.31) and (2.32), we obtain V^m and corresponding optimal mixed strategy $\boldsymbol{y}^{R*}(1) = (y_1^{R*}(1), y_2^{R*}(1))^{\mathrm{T}}$ for the player I, where

$$\begin{aligned} V^m &= v^R(1) = \frac{1}{\frac{1}{204} + \frac{1}{765}} = \frac{52020}{323} \approx 161.0526, \\ y_1^{R*}(1) &= \frac{52020}{323} \times \frac{1}{204} = \frac{255}{323} \approx 0.7895 \end{aligned}$$

and

$$y_2^{R*}(1) = \frac{52020}{323} \times \frac{1}{765} = \frac{68}{323} \approx 0.2105.$$

Analogously, according to Eq. (2.55), the linear programming model is constructed as follows:

$$\min\{x_1^R(0) + x_2^R(0)\}$$
$$\text{s.t.} \begin{cases} 190x_1^R(0) + 100x_2^R(0) \geq 1 \\ 158x_1^R(0) + 190x_2^R(0) \geq 1 \\ x_1^R(0) \geq 0,\ x_2^R(0) \geq 0, \end{cases}$$

where $x_1^R(0)$ and $x_2^R(0)$ are decision variables. Solving the above linear programming model, we obtain its optimal solution $\boldsymbol{x}^{R*}(0) = (x_1^{R*}(0), x_2^{R*}(0))^{\mathrm{T}}$, where

$$x_1^{R*}(0) = \frac{9}{2030} \approx 0.0044, \quad x_2^{R*}(0) = \frac{8}{5075} \approx 0.0016.$$

According to Eqs. (2.31) and (2.32), we obtain V^r and corresponding optimal mixed strategy $\boldsymbol{y}^{R*}(0) = (y_1^{R*}(0), y_2^{R*}(0))^{\mathrm{T}}$ for the player I, where

$$\begin{aligned} V^r &= v^R(0) = \frac{1}{\frac{9}{2030} + \frac{8}{5075}} = \frac{2060450}{12383} \approx 166.3934, \\ y_1^{R*}(0) &= \frac{2060450}{12383} \times \frac{9}{2030} = \frac{9135}{12383} \approx 0.7377 \end{aligned}$$

and

$$y_2^{R*}(0) = \frac{2060450}{12383} \times \frac{8}{5075} = \frac{3248}{12383} \approx 0.2623.$$

According to Eq. (2.57), the linear programming model is constructed as follows:

$$\min\{x_1^L(0)+x_2^L(0)\}$$
$$\text{s.t.}\begin{cases}175x_1^L(0)+80x_2^L(0)\geq 1\\ 150x_1^L(0)+175x_2^L(0)\geq 1\\ x_1^L(0)\geq 0, x_2^L(0)\geq 0,\end{cases}$$

where $x_1^L(0)$ and $x_2^L(0)$ are decision variables. Solving the above linear programming model, we obtain its optimal solution $\boldsymbol{x}^{L*}(0)=(x_1^{L*}(0),x_2^{L*}(0))^{\mathrm{T}}$, where

$$x_1^{L*}(0)=\frac{19}{3725}\approx 0.0051,\quad x_2^{L*}(0)=\frac{1}{745}\approx 0.0013.$$

According to Eqs. (2.37) and (2.38), we obtain V^l and the optimal mixed strategy $\boldsymbol{y}^{L*}(0)=(y_1^{L*}(0),y_2^{L*}(0))^{\mathrm{T}}$ for the player I, where

$$V^l=v^L(0)=\frac{1}{\frac{19}{3725}+\frac{1}{745}}=\frac{3725}{24}\approx 155.2083,$$
$$y_1^{L*}(0)=\frac{3725}{24}\times\frac{19}{3725}=\frac{19}{24}\approx 0.7917$$

and

$$y_2^{L*}(0)=\frac{3725}{24}\times\frac{1}{745}=\frac{5}{24}\approx 0.2083.$$

Therefore, the fuzzy value of the matrix game $\tilde{\boldsymbol{A}}_2$ with payoffs of triangular fuzzy numbers can be directly obtained as $\tilde{V}^{\prime *}=(V^l,V^m,V^r)=(155.2083,161.0526,166.3934)$, whose membership function is given as follows:

$$\mu_{\tilde{V}^{\prime *}}(x)=\begin{cases}\frac{x-155.2083}{5.8443} & \text{if } 155.2083\leq x<161.0526\\ 1 & \text{if } x=161.0526\\ \frac{166.3934-x}{5.3408} & \text{if } 161.0526<x\leq 166.3934\\ 0 & \text{else,}\end{cases}$$

depicted as in Fig. 2.8.

2. Computational results obtained by other methods and analysis

The above numerical example was solved by the two-level linear programming method proposed in Sect. 2.4 and the lexicographic method proposed in Sect. 2.5.

Fig. 2.8 The fuzzy value $\tilde{V}'^{*}$

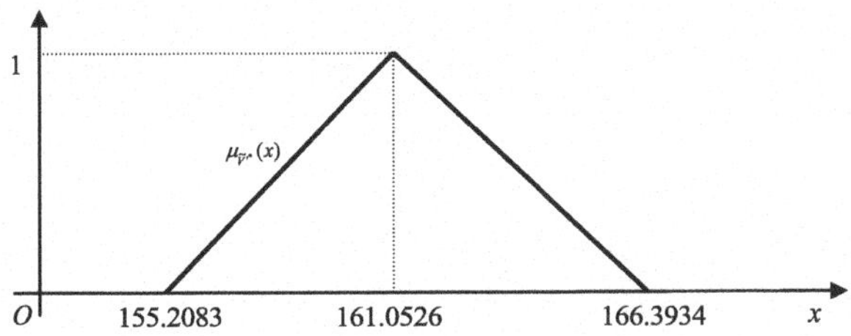

In this subsection, this matrix game $\tilde{\boldsymbol{A}}_2$ with payoffs of triangular fuzzy numbers is solved by other methods [7, 14]. The computational results are analyzed and compared to show the validity, applicability, and superiority of the proposed method in this Section.

(2a) Computational results obtained by Campos' method

Taking the players' gain-floor and loss-ceiling as crisp values, following a similar way to crisp matrix games [4, 26], i.e., according to Eqs. (2.8) and (2.9), using a suitable defuzzification (i.e., linear ranking) function of fuzzy numbers, Campos [7] constructed the auxiliary linear programming models as follows:

$$\min\left\{\sum_{i=1}^{m} u_i^C\right\}$$
$$\text{s.t.}\begin{cases}\sum\limits_{i=1}^{m}(a_{ij}^l + a_{ij}^m + a_{ij}^r)u_i^C \geq 3 - (1-\lambda)(p_j^l + p_j^m + p_j^r) \quad (j = 1, 2, \ldots, n)\\ u_i^C \geq 0 \quad (i = 1, 2, \ldots, m)\end{cases} \tag{2.59}$$

and

$$\max\left\{\sum_{j=1}^{n} v_j^C\right\}$$
$$\text{s.t.}\begin{cases}\sum\limits_{j=1}^{n}(a_{ij}^l + a_{ij}^m + a_{ij}^r)v_j^C \leq 3 + (1-\tau)(q_i^l + q_i^m + q_i^r) \quad (i = 1, 2, \ldots, m)\\ v_j^C \geq 0 \quad (j = 1, 2, \ldots, n),\end{cases} \tag{2.60}$$

where $\lambda \in [0,1]$ and $\tau \in [0,1]$, $\tilde{p}_j = (p_j^l, p_j^m, p_j^r)$ and $\tilde{q}_i = (q_i^l, q_i^m, q_i^r)$ are triangular fuzzy numbers, and

$$u_i^C = \frac{y_i^C}{v^C} \quad (i = 1, 2, \ldots, m) \tag{2.61}$$

and

$$v_j^C = \frac{z_j^C}{\omega^C} \quad (j = 1, 2, \ldots, n) \tag{2.62}$$

are decision variables.

For the aforementioned matrix game $\tilde{A}_2$ with payoffs of triangular fuzzy numbers, according to Eqs. (2.59) and (2.60), the linear programming models are constructed as follows:

$$\min\{u_1^C + u_2^C\}$$
$$\text{s.t.} \begin{cases} 545u_1^C + 270u_2^C \geq 3 - 0.29(1-\lambda) \\ 464u_1^C + 545u_2^C \geq 3 - 0.29(1-\lambda) \\ u_1^C \geq 0, u_2^C \geq 0 \end{cases}$$

and

$$\max\{v_1^C + v_2^C\}$$
$$\text{s.t.} \begin{cases} 545v_1^C + 464v_2^C \leq 3 + 0.46(1-\tau) \\ 270v_1^C + 545v_2^C \leq 3 + 0.46(1-\tau) \\ v_1^C \geq 0, v_2^C \geq 0, \end{cases}$$

where $\tilde{p}_1 = \tilde{p}_2 = (0.08, 0.10, 0.11)$ and $\tilde{q}_1 = \tilde{q}_2 = (0.14, 0.15, 0.17)$ are taken from Campos [7].

Solving the above linear programming models by using the simplex method of linear programming, and combining with Eqs. (2.61) and (2.62), we obtain the player I's gain-floor and the player II' loss-ceiling and their optimal mixed strategies as follows:

$$\begin{aligned} v^{*C}(\lambda) &= \frac{171745}{356[3 - 0.29(1-\lambda)]} \approx \frac{160.8099}{1 - 0.0967(1-\lambda)}, \\ (y_1^{*C}, y_2^{*C})^{\mathrm{T}} &= (\frac{275}{356}, \frac{81}{356})^{\mathrm{T}} \approx (0.7725, 0.2275)^{\mathrm{T}}, \\ \omega^{*C}(\tau) &= \frac{171745}{356[3 + 0.46(1-\tau)]} \approx \frac{160.8099}{1 + 0.1533(1-\tau)} \end{aligned}$$

and

$$(z_1^{*C}, z_2^{*C})^{\mathrm{T}} = (\frac{81}{356}, \frac{275}{356})^{\mathrm{T}} \approx (0.2275, 0.7725)^{\mathrm{T}},$$

respectively. Obviously, $v^{*C}(\lambda) \geq \omega^{*C}(\tau)$. Moreover, $\omega^{*C}(\tau)$ is an increasing function of $\tau \in [0, 1]$ whereas $v^{*C}(\lambda)$ is a decreasing function of $\lambda \in [0, 1]$. It easily follows that $v^{*C}(1) = \omega^{*C}(1) = 160.8099$ when $\lambda = \tau = 1$. Thus, Campos [7] argued that the matrix game $\tilde{A}_2$ with payoffs of triangular fuzzy numbers has the fuzzy value "close to 160.8099".

(2b) Computational results obtained by Bector et al.'s method

Taking the players' gain-floor and loss-ceiling as fuzzy numbers, using a suitable defuzzification function F, according to Eqs. (2.8) and (2.9) and the concept of double fuzzy constraints [7], Bector et al. [14] (with reference to [12, 13]) suggested the mathematical programming models for the players I and II as follows:

$$\max\{F(\tilde{v}^B)\}$$
$$\text{s.t.}\begin{cases} \sum_{i=1}^{m} F(\tilde{a}_{ij}) y_i^B \geq F(\tilde{v}^B) - (1-\lambda) F(\tilde{p}_j) \quad (j = 1, 2, \ldots, n) \\ \sum_{i=1}^{m} y_i^B = 1 \\ y_i^B \geq 0 \quad (i = 1, 2, \ldots, m) \end{cases} \tag{2.63}$$

and

$$\min\{F(\tilde{\omega}^B)\}$$
$$\text{s.t.}\begin{cases} \sum_{j=1}^{n} F(\tilde{a}_{ij}) z_j^B \leq F(\tilde{\omega}^B) + (1-\tau) F(\tilde{q}_i) \quad (i = 1, 2, \ldots, m) \\ \sum_{j=}^{n} z_j^B = 1 \\ z_j^B \geq 0 \quad (j = 1, 2, \ldots, n), \end{cases} \tag{2.64}$$

respectively, where $\tilde{p}_j$ and $\tilde{q}_i$ $(i = 1, 2, \ldots, m;\ j = 1, 2, \ldots, n)$ are fuzzy numbers, $\lambda \in [0, 1]$, $\tau \in [0, 1]$.

In the case that $\tilde{A}$ is the matrix game with payoffs of triangular fuzzy numbers, i.e., all $\tilde{v}^B = (v^{Bl}, v^{Bm}, v^{Br})$, $\tilde{\omega}^B = (\omega^{Bl}, \omega^{Bm}, \omega^{Br})$, $\tilde{a}_{ij} = (a_{ij}^l, a_{ij}^m, a_{ij}^r)$, $\tilde{p}_j = (p_j^l, p_j^m, p_j^r)$, and $\tilde{q}_i = (q_i^l, q_i^m, q_i^r)$ $(i = 1, 2, \ldots, m; j = 1, 2, \ldots, n)$ are triangular fuzzy numbers, using Yager's index [27], Bector et al. [14] transformed Eqs. (2.63) and (2.64) into the following linear programming models:

$$\max\{v^B\}$$
$$\text{s.t.}\begin{cases}\sum_{i=1}^{m}(a_{ij}^l+a_{ij}^m+a_{ij}^r)y_i^B \geq 3v^B-(1-\lambda)(p_j^l+p_j^m+p_j^r) \quad (j=1,2,\ldots,n)\\ \sum_{i=1}^{m} y_i^B=1\\ y_i^B\geq 0 \quad (i=1,2,\ldots,m)\end{cases} \tag{2.65}$$

and

$$\min\{\omega^B\}$$
$$\text{s.t.}\begin{cases}\sum_{j=1}^{n}(a_{ij}^l+a_{ij}^m+a_{ij}^r)z_j^B \leq 3\omega^B+(1-\tau)(q_i^l+q_i^m+q_i^r) \quad (i=1,2,\ldots,m)\\ \sum_{j=}^{n} z_j^B=1\\ z_j^B\geq 0 \quad (j=1,2,\ldots,n),\end{cases} \tag{2.66}$$

respectively, where

$$v^B=F(\tilde{v}^B)=\frac{v^{Bl}+v^{Bm}+v^{Br}}{3} \tag{2.67}$$

and

$$\omega^B=F(\tilde{\omega}^B)=\frac{\omega^{Bl}+\omega^{Bm}+\omega^{Br}}{3}. \tag{2.68}$$

For the aforementioned matrix game $\tilde{A}_2$ with payoffs of triangular fuzzy numbers, according to Eqs. (2.65) and (2.66) with $\tilde{p}_1=\tilde{p}_2=(0.08,0.10,0.11)$ and $\tilde{q}_1=\tilde{q}_2=(0.14,0.15,0.17)$, the linear programming models are constructed as follows:

$$\max\{v^B\}$$
$$\text{s.t.}\begin{cases}545y_1^B+270y_2^B\geq 3v^B-0.29(1-\lambda)\\ 464y_1^B+545y_2^B\geq 3v^B-0.29(1-\lambda)\\ y_1^B+y_2^B=1\\ y_1^B\geq 0, y_2^B\geq 0\end{cases}$$

and

$$\min\{\omega^B\}$$
$$\text{s.t.}\begin{cases} 545z_1^B + 464z_2^B \leq 3\omega^B + 0.46(1-\tau) \\ 270z_1^B + 545z_2^B \leq 3\omega^B + 0.46(1-\tau) \\ z_1^B + z_2^B = 1 \\ z_1^B \geq 0, z_2^B \geq 0, \end{cases}$$

respectively. Simply computing/solving the above linear programming models, we can obtain the player I's gain-floor, the player II' loss-ceiling, and their optimal mixed strategies as follows:

$$\begin{aligned} v^{*B}(\lambda) &= \frac{171745}{1068} + \frac{0.29(1-\lambda)}{3} \approx 160.8099 + 0.0967(1-\lambda), \\ (y_1^{*B}, y_2^{*B})^{\mathrm{T}} &= (\frac{275}{356}, \frac{81}{356})^{\mathrm{T}} = (0.7725, 0.2275)^{\mathrm{T}}, \\ \omega^{*B}(\tau) &= \frac{171745}{1068} - \frac{0.46(1-\tau)}{3} \approx 160.8099 - 0.1533(1-\tau) \end{aligned}$$

and

$$(z_1^{*B}, z_2^{*B})^{\mathrm{T}} = (\frac{81}{356}, \frac{275}{356})^{\mathrm{T}} = (0.2275, 0.7725)^{\mathrm{T}},$$

respectively.

Obviously, $v^{*B}(\lambda)$ and $\omega^{*B}(\tau)$ remarkably differ from $v^{*C}(\lambda)$ and $\omega^{*C}(\tau)$ when $\lambda \neq 1$ and $\tau \neq 1$.

3. Computational result comparison and conclusions

Comparing the aforementioned modeling, methods, and computational results, we can easily draw the following conclusions.

(3a) Modeling. The players' gain-floor and loss-ceiling were regarded as triangular fuzzy numbers in the proposed methods in this section and Sects. 2.4, 2.5 and Bector et al.'s [14]. However, they were regarded as real numbers in Campos's method [7]. This case is not rational since the players' expected payoffs are a linear combination of fuzzy payoffs which are expressed with triangular fuzzy numbers.

(3b) Process and methods. The proposed method in this section is developed on the monotonicity of values of matrix games. It always ensures that any matrix game with payoffs of triangular fuzzy numbers has a fuzzy value, which is a triangular fuzzy number also. Moreover, the fuzzy value can be directly and explicitly obtained by solving the derived three linear programming models with data taken from 1-cut set and 0-cut set of fuzzy payoffs. Li's model as stated in Sect. 2.4 was developed on the ordering relation of triangular fuzzy numbers [20] and multi-objective programming. The derived six linear programming models were used to compute the players' gain-floor and loss-ceiling. Obviously, Li's model in Sect. 2.4 depended on the ordering relation. Following a similar way to crisp matrix games, based on the concept of double fuzzy constraints and ranking functions, Campos's method [7] regarded the players' gain-floor and loss-ceiling as real numbers and hereby suggested two auxiliary linear programming models. Bector et al.'s method [14] was developed on certain duality of linear programming with fuzzy parameters. As Bector et al. [12] themselves pointed out, Bector et al.'s method [14] was essentially the same as that of Campos [7]. Campos's method and Bector et al.'s method are defuzzification approaches, which not only closely depend on ranking functions, parameters, and adequacies but also cannot explicitly obtain membership functions of the players' gain-floor and loss-ceiling.

(3c) Computational results. The proposed method in this section can explicitly obtain the fuzzy value $\tilde{V}'^* = (155.2083, 161.0526, 166.3934)$ of the matrix game $\tilde{\boldsymbol{A}}_2$ with payoffs of triangular fuzzy numbers. Li's model in Sect. 2.4 can explicitly obtain the player I's gain-floor $\tilde{\upsilon}^* = (154.9996, 161.05, 164.737)$ and the player II's loss-ceiling $\tilde{\omega}^* = (155.2633, 161.05, 171.0523)$ which are not identical. This case is not rational since the matrix game is zero-sum. Moreover, it is intuitively seen from Fig. 2.6 that $\tilde{V}^* = (155.2633, 161.05, 164.737)$ is better than $\tilde{\upsilon}^*$ and $\tilde{\omega}^*$. In fact, using Yager's index F [27], i.e., Eq. (2.67) or Eq. (2.68), we have

$$F(\tilde{\upsilon}^*) = \frac{154.9996 + 161.05 + 164.737}{3} = 160.2622,$$
$$F(\tilde{V}^*) = \frac{155.2633 + 161.05 + 164.737}{3} = 160.3501$$

and

$$F(\tilde{\omega}^*) = \frac{155.2633 + 161.05 + 171.0523}{3} = 162.4552,$$

which infers that $F(\tilde{\upsilon}^*) < F(\tilde{V}^*) < F(\tilde{\omega}^*)$. Therefore, $\tilde{\upsilon}^* < \tilde{V}^* < \tilde{\omega}^*$.

Campos's method [7] provided crisp values for the players' gain-floor and loss-ceiling in the matrix game $\tilde{\boldsymbol{A}}_2$ with payoffs of triangular fuzzy numbers. Bector et al.'s method [14] provided defuzzification values of the players' gain-floor and loss-ceiling. Namely, these two methods cannot explicitly obtain membership functions of the players' gain-floor and loss-ceiling even though these are very much desirable. Moreover, these methods cannot always guarantee that the defuzzification values are identical and the matrix game $\tilde{A}_2$ with payoffs of triangular fuzzy numbers has a defuzzification value. On the other hand, the defuzzification values closely depend on not only choice of ranking functions but also the parameters and adequacies, which are difficult to be appropriately determined a priori.

(3d) Computational complexity. The proposed method in this section needs to solve three linear programming models. Li's model proposed in Sect. 2.4 needs to solve six linear programming models with additional decision variables and constraints, which usually may be superabundant and even contradictable. However, Campos's method [7] and Bector et al.'s method [14] need to solve a series of linear programming models for different parameters and adequacies. Therefore, the computational amount and complexity of the proposed method in this section are less than those of Li's model, Campos's method, and Bector et al.'s method.

References

1. Butnariu D (1978) Fuzzy games: a description of the concept. Fuzzy Sets Syst 1:181–192
2. Aubin JP (1981) Cooperative fuzzy game. Math Oper Res 6:1–13
3. Dubois D, Prade H (1980) Fuzzy sets and systems: theory and applications. Academic Press, New York
4. Li D-F (2003) Fuzzy multiobjective many-person decision makings and games. National Defense Industry Press, Beijing (in Chinese)
5. Zadeh L (1965) Fuzzy sets. Inf Control 8:338–356
6. Li D-F (2014) Decision and game theory in management with intuitionistic fuzzy sets. Springer, Heidelberg
7. Campos L (1989) Fuzzy linear programming models to solve fuzzy matrix games. Fuzzy Sets Syst 32:275–289
8. Campos L, Gonzalez A (1991) Fuzzy matrix games considering the criteria of the players. Kybernetes 20:17–23
9. Campos L, Gonzalez A, Vila MA (1992) On the use of the ranking function approach to solve fuzzy matrix games in a direct way. Fuzzy Sets Syst 49:193–203
10. Nishizaki I, Sakawa M (2001) Fuzzy and multiobjective games for conflict resolution. Springer, Physica-Verlag, Berlin
11. Sakawa M, Nishizaki I (1994) Max-min solutions for fuzzy multiobjective matrix games. Fuzzy Sets Syst 67:53–69
12. Bector CR, Chandra S (2005) Fuzzy mathematical programming and fuzzy matrix games. Springer, Berlin
13. Bector CR, Chandra S, Vijay V (2004) Matrix games with fuzzy goals and fuzzy linear programming duality. Fuzzy Optim Decis Making 3:255–269
14. Bector CR, Chandra S, Vijay V (2004) Duality in linear programming with fuzzy parameters and matrix games with fuzzy pay-offs. Fuzzy Sets Syst 46(2):253–269
15. Vijay V, Chandra S, Bector CR (2005) Matrix games with fuzzy goals and fuzzy payoffs. Omega: Int J Manag Sci 33:425–429
16. Li D-F (1999) A fuzzy multiobjective programming approach to solve fuzzy matrix games. J Fuzzy Math 7(4):907–912

17. Li D-F, Yang J-B (2004) Two level linear programming approach to solve fuzzy matrix games with fuzzy payoffs. University of Manchester Institute of Science and Technology, Manchester School of Management, UK, Unpublished preprint
18. Larbani M (2009) Non cooperative fuzzy games in normal form: a survey. Fuzzy Sets Syst 160:3184–3210
19. Moore RE (1979) Method and application of interval analysis. SIAM, Philadelphia
20. Ramik J, Rimanek J (1985) Inequality relation between fuzzy numbers and its use in fuzzy optimization. Fuzzy Sets Syst 16:123–138
21. Chankong V, Haimes YY (1983) Multiobjective decision making: theory and methodology. North-Holland, New York
22. Steuer RE (1986) Multiple criteria optimization: theory, computation, and application. Wiley, New York
23. Li D-F, Cheng C-T (2002) Fuzzy multiobjective programming methods for fuzzy constrained matrix games with fuzzy numbers. Int J Uncertainty Fuzziness Knowl Based Syst 10(4):385–400
24. Zimmermann H-J (1991) Fuzzy set theory and its application, 2nd edn. Kluwer Academic Publishers, Dordrecht
25. Li D-F (1999) Fuzzy constrained matrix games with fuzzy payoffs. J Fuzzy Math 7(4):873–880
26. Owen G (1982) Game theory, 2nd edn. Academic Press, New York
27. Yager RR (1981) A procedure for ordering fuzzy numbers of the unit interval. Inf Sci 24:143–161

Part II
Models and Methods of Constrained Matrix Games with Payoffs of Triangular Fuzzy Numbers

Chapter 3
Interval-Valued Constrained Matrix Games

3.1 Introduction

As stated in Chaps. 1 and 2, many real-life competitive and conflict decision problems can be modeled as interval-valued or fuzzy matrix games [1–4]. In these matrix games, the players can arbitrary choose their strategies. On other words, choice of strategies for the players is not constrained. However, in some real-life game problems, choice of strategies for the players is constrained due to some practical reason why this should be (see Chap. 5 of Dresher [5] and Page 58–59 of Owen [1] for references), i.e., not all mixed (or pure) strategies in a game are permitted for each player [6]. Such a two-person zero-sum finite game is called a matrix game with sets of constraint strategies, which often is called as a constrained matrix game for short. Dresher [5] gave a real example of the constrained matrix game. Li and Cheng [7] studied a constrained matrix game with fuzzy payoffs, which is called a fuzzy constrained matrix game. In most of the fuzzy matrix games, the payoffs were expressed with fuzzy numbers whose membership functions are already known a priori. These membership functions play an important role in corresponding methods. In reality, it is not always easy for the players to specify the membership functions in uncertain environments. In some cases, the payoffs are easily estimated as intervals [8]. As far as we know, no studies have yet been attempted for interval-valued matrix games with sets of constraint strategies, which often are called interval-valued constrained matrix games for short. Thus, in this chapter, we focus on studying interval-valued constrained matrix games. In Sect. 3.2, we briefly review the definitions and notations of constrained matrix games. In Sect. 3.3, we formulate interval-valued constrained matrix games and discuss their important properties and hereby develop a primal-dual linear programming method for solving interval-valued constrained matrix games.

D.-F. Li, *Linear Programming Models and Methods of Matrix Games with Payoffs of Triangular Fuzzy Numbers*, Studies in Fuzziness and Soft Computing 328 ,
DOI 10.1007/978-3-662-48476-0_3

3.2 Constrained Matrix Games and Auxiliary Linear Programming Models

Constrained matrix games are matrix games in which not all mixed (or pure) strategies are permitted for each player. More precisely, a constrained matrix game is described as follows. Assume that $S_1 = \{\delta_1, \delta_2, \ldots, \delta_m\}$ and $S_2 = \{\beta_1, \beta_2, \ldots, \beta_n\}$ are sets of pure strategies for the players I and II, respectively. A payoff matrix of the player I is $\boldsymbol{A}$ as stated in Sect. 1.2, i.e.,

$$\boldsymbol{A} = (a_{ij})_{m\times n} = \begin{array}{c} \\ \delta_1 \\ \delta_2 \\ \vdots \\ \delta_m \end{array} \begin{array}{c} \begin{array}{cccc} \beta_1 & \beta_2 & \cdots & \beta_n \end{array} \\ \begin{pmatrix} a_{11} & a_{12} & \cdots & a_{1n} \\ a_{21} & a_{22} & \cdots & a_{2n} \\ \vdots & \vdots & \cdots & \vdots \\ a_{m1} & a_{m2} & \cdots & a_{mn} \end{pmatrix} \end{array}.$$

The mixed strategies $\boldsymbol{y} = (y_1, y_2, \ldots, y_m)^{\mathrm{T}}$ and $\boldsymbol{z} = (z_1, z_2, \ldots, z_n)^{\mathrm{T}}$ respectively must be chosen from some convex hyper-polyhedron, i.e., from the constraint sets determined by the systems of linear inequalities and/or equations. Without loss of generality, let $Y = \{\boldsymbol{y} | \boldsymbol{B}^{\mathrm{T}}\boldsymbol{y} \leq \boldsymbol{c}, \boldsymbol{y} \geq 0\}$ represent the constraint set of mixed strategies for the player I, where $\boldsymbol{c} = (c_1, c_2, \ldots, c_p)^{\mathrm{T}}$, $\boldsymbol{B} = (b_{il})_{m\times p}$, p is a positive integer and $\boldsymbol{0}$ is an adequate-dimensional vector in which all components are 0. The constraint of $\boldsymbol{y}$ satisfying $\sum_{i=1}^{m} y_i = 1$ as a mixed strategy is expressed in the system of inequalities: $\boldsymbol{B}^{\mathrm{T}}\boldsymbol{y} \leq \boldsymbol{c}$. Analogously, let $Z = \{\boldsymbol{z} | \boldsymbol{E}\boldsymbol{z} \geq \boldsymbol{d}, \boldsymbol{z} \geq 0\}$ denote the constraint set of mixed strategies for the player II, where $\boldsymbol{d} = (d_1, d_2, \ldots, d_q)^{\mathrm{T}}$, $\boldsymbol{E} = (e_{kj})_{q\times n}$ and q is a positive integer. The constraint of $\boldsymbol{z}$ satisfying $\sum_{i=1}^{n} z_i = 1$ as a mixed strategy is expressed in the system of inequalities: $\boldsymbol{E}\boldsymbol{z} \geq \boldsymbol{d}$. In a parallel way to matrix games as stated in Sect. 1.2, the player I should choose an optimal (mixed) strategy $\boldsymbol{y}^* \in Y$ so that

$$\min_{z\in Z}\{\boldsymbol{y}^{*\mathrm{T}}\boldsymbol{A}\boldsymbol{z}\} = \max_{y\in Y}\min_{z\in Z}\{\boldsymbol{y}^{\mathrm{T}}\boldsymbol{A}\boldsymbol{z}\}. \tag{3.1}$$

Similarly, the player II should choose an optimal (mixed) strategy $\boldsymbol{z}^* \in Z$ so as to obtain

$$\max_{y\in Y}\{\boldsymbol{y}^{\mathrm{T}}\boldsymbol{A}\boldsymbol{z}^*\} = \min_{z\in Z}\max_{y\in Y}\{\boldsymbol{y}^{\mathrm{T}}\boldsymbol{A}\boldsymbol{z}\}. \tag{3.2}$$

Definition 3.1 If there exist $\boldsymbol{y}^* \in Y$ and $\boldsymbol{z}^* \in Z$ so that $(\boldsymbol{y}^*, \boldsymbol{z}^*)$ satisfies

$$\boldsymbol{y}^{*\mathrm{T}}\boldsymbol{A}\boldsymbol{z}^* = \min_{z\in Z}\{\boldsymbol{y}^{*\mathrm{T}}\boldsymbol{A}\boldsymbol{z}\} = \max_{y\in Y}\{\boldsymbol{y}^{\mathrm{T}}\boldsymbol{A}\boldsymbol{z}^*\}$$

for all mixed strategies $\boldsymbol{y} \in Y$ and $\boldsymbol{z} \in Z$, then $(\boldsymbol{y}^*, \boldsymbol{z}^*)$ and $V = \boldsymbol{y}^{*\mathrm{T}}\boldsymbol{A}\boldsymbol{z}^*$ are called a saddle point in the sense of mixed strategies and a value of the constrained matrix game $\boldsymbol{A}$, respectively.

Theorem 3.1 *If there exists* $(\boldsymbol{y}^*, \boldsymbol{z}^*)$, *where* $\boldsymbol{y}^* \in Y$ *and* $\boldsymbol{z}^* \in Z$, *so that*

$$\boldsymbol{y}^{\mathrm{T}}\boldsymbol{A}\boldsymbol{z}^* \le \boldsymbol{y}^{*\mathrm{T}}\boldsymbol{A}\boldsymbol{z}^* \le \boldsymbol{y}^{*\mathrm{T}}\boldsymbol{A}\boldsymbol{z} \tag{3.3}$$

for all mixed strategies $\boldsymbol{y} \in Y$ *and* $\boldsymbol{z} \in Z$, *then* $(\boldsymbol{y}^*, \boldsymbol{z}^*)$ *and* $\mathrm{V} = \boldsymbol{y}^{*\mathrm{T}}\boldsymbol{A}\boldsymbol{z}^*$ *are a saddle point and a value of the constrained matrix game* $\boldsymbol{A}$, *respectively.*

Proof It follows from Eq. (3.3) that

$$\max_{y \in Y}\{\boldsymbol{y}^{\mathrm{T}}\boldsymbol{A}\boldsymbol{z}^*\} \le \boldsymbol{y}^{*\mathrm{T}}\boldsymbol{A}\boldsymbol{z}^* \le \min_{z \in Z}\{\boldsymbol{y}^{*\mathrm{T}}\boldsymbol{A}\boldsymbol{z}\},$$

which infers that

$$\min_{z \in Z}\max_{y \in Y}\{\boldsymbol{y}^{\mathrm{T}}\boldsymbol{A}\boldsymbol{z}\} \le \boldsymbol{y}^{*\mathrm{T}}\boldsymbol{A}\boldsymbol{z}^* \le \max_{y \in Y}\min_{z \in Z}\{\boldsymbol{y}^{\mathrm{T}}\boldsymbol{A}\boldsymbol{z}\}. \tag{3.4}$$

On the other hand, for all mixed strategies $\boldsymbol{y} \in Y$ and $\boldsymbol{z} \in Z$, we obtain

$$\boldsymbol{y}^{\mathrm{T}}\boldsymbol{A}\boldsymbol{z} \le \max_{y \in Y}\{\boldsymbol{y}^{\mathrm{T}}\boldsymbol{A}\boldsymbol{z}\}.$$

Furthermore, we have

$$\min_{z \in Z}\{\boldsymbol{y}^{\mathrm{T}}\boldsymbol{A}\boldsymbol{z}\} \le \min_{z \in Z}\max_{y \in Y}\{\boldsymbol{y}^{\mathrm{T}}\boldsymbol{A}\boldsymbol{z}\},$$

which infers that

$$\max_{y \in Y}\min_{z \in Z}\{\boldsymbol{y}^{\mathrm{T}}\boldsymbol{A}\boldsymbol{z}\} \le \min_{z \in Z}\max_{y \in Y}\{\boldsymbol{y}^{\mathrm{T}}\boldsymbol{A}\boldsymbol{z}\}.$$

Combining with both Eq. (3.4) and Definition 3.1, we can prove that the conclusion is true and the proof has been completed.

In fact, Theorem 3.1 presents a necessary and sufficient condition for existence of saddle points of constrained matrix games. But, Theorem 3.1 is not a very efficient tool. In the following, we can characterize the conclusion of Theorem 3.1 with rather elegant concepts of both a linear programming and its duality. By using the duality theorem of linear programming [1], Eqs. (3.1) and (3.2) are equivalent to the linear programming models as follows:

$$\max\{\boldsymbol{d}^{\mathrm{T}}\boldsymbol{x}\}$$
$$\text{s.t.}\begin{cases} \boldsymbol{E}^{\mathrm{T}}\boldsymbol{x} - \boldsymbol{A}^{\mathrm{T}}\boldsymbol{y} \le 0 \\ \boldsymbol{B}^{\mathrm{T}}\boldsymbol{y} \le \boldsymbol{c} \\ \boldsymbol{x} \ge 0 \\ \boldsymbol{y} \ge 0, \end{cases} \tag{3.5}$$

and

$$\min\{\boldsymbol{c}^{\mathrm{T}}\boldsymbol{s}\}$$
$$\text{s.t}\begin{cases}\boldsymbol{Bs}-\boldsymbol{Az}\geq 0\\ \boldsymbol{Ez}\geq\boldsymbol{d}\\ \boldsymbol{s}\geq 0\\ \boldsymbol{z}\geq 0,\end{cases}\tag{3.6}$$

respectively, where $\boldsymbol{x}=(x_1,x_2,\ldots,x_q)^{\mathrm{T}}$ and $\boldsymbol{s}=(s_1,s_2,\ldots,s_p)^{\mathrm{T}}$.

It is easy to see that Eqs. (3.5) and (3.6) are a pair of primal-dual linear programming models. Therefore, Eqs. (3.1) and (3.2) are equal if both Eqs. (3.5) and (3.6) are feasible. Thus, the constrained matrix game $\boldsymbol{A}$ must have a saddle point in the sense of mixed strategies. We draw a conclusion as follow.

Theorem 3.2 If Eqs. (3.5) *and* (3.6) *are feasible linear programming, then they have optimal solutions* $(\boldsymbol{y}^*,\boldsymbol{x}^*)^{\mathrm{T}}$ and $(\boldsymbol{z}^*,\boldsymbol{s}^*)^{\mathrm{T}}$, *respectively* [1]. *Moreover,* $(\boldsymbol{y}^*,\boldsymbol{z}^*)$ and $V=\boldsymbol{y}^{*\mathrm{T}}\boldsymbol{A}\boldsymbol{z}^*$ *are a saddle point and a value of the constrained matrix game* $\boldsymbol{A}$, *respectively.*

Theorem 3.3 *If* $(\boldsymbol{y}^*,\boldsymbol{x}^*)^{\mathrm{T}}$ *and* $(\boldsymbol{z}^*,\boldsymbol{s}^*)^{\mathrm{T}}$ *are feasible solutions of* Eqs. (3.5) *and* (3.6), *respectively, and* $\boldsymbol{d}^{\mathrm{T}}\boldsymbol{x}^*=\boldsymbol{c}^{\mathrm{T}}\boldsymbol{s}^*$, *then* $(\boldsymbol{y}^*,\boldsymbol{z}^*)$ *and* $V=\boldsymbol{y}^{*\mathrm{T}}\boldsymbol{A}\boldsymbol{z}^*$ *are a saddle point and a value of the constrained matrix game* $\boldsymbol{A}$, *respectively.*

Proof By using the duality theorem of linear programming, we can easily prove Theorem 3.3 (omitted).

However, Eqs. (3.5) and (3.6) need not always have feasible solutions. As a result, not all constrained matrix games have saddle points in the sense of mixed strategies.

3.3 Primal-Dual Linear Programming Models of Interval-Valued Constrained Matrix Games

Let us consider the following interval-valued constrained matrix game, where the sets S_1 and S_2 of pure strategies and the constraint sets Y and Z of mixed strategies for the players I and II are defined as in Sect. 3.2, the interval-valued payoff matrix $\bar{\boldsymbol{A}}$ of the player I is defined as in Sect. 1.3.2, i.e.,

$$\bar{\boldsymbol{A}}=(\bar{a}_{ij})_{m\times n}=\begin{matrix}\\ \delta_1\\ \delta_2\\ \vdots\\ \delta_m\end{matrix}\begin{matrix}\beta_1 & \beta_2 & \cdots & \beta_n\\ \left(\begin{matrix}[a_{L11},a_{R11}] & [a_{L12},a_{R12}] & \cdots & [a_{L1n},a_{R1n}]\\ [a_{L21},a_{R21}] & [a_{L22},a_{R22}] & \cdots & [a_{L2n},a_{R2n}]\\ \vdots & \vdots & \cdots & \vdots\\ [a_{Lm1},a_{Rm1}] & [a_{Lm2},a_{Rm2}] & \cdots & [a_{Lmn},a_{Rmn}]\end{matrix}\right)\end{matrix}.$$

In the sequent, such a matrix game is often called the interval-valued constrained matrix game $\bar{\boldsymbol{A}}$.

3.3.1 Monotonicity of Values of Constrained Matrix Games

For any given values a_{ij} in the interval-valued payoffs $\bar{a}_{ij} = [a_{Lij}, a_{Rij}]$ $(i = 1, 2, \ldots, m;\ j = 1, 2, \ldots, n)$, a payoff matrix is denoted by $\boldsymbol{A} = (a_{ij})_{m\times n}$. Thus, we may construct a constrained matrix game $\boldsymbol{A}$ whose payoff matrix of the player I is $\boldsymbol{A} = (a_{ij})_{m\times n}$, where the constraint sets Y and Z of mixed strategies for the players I and II are defined as in Sect. 3.2, respectively.

It is easy to see from Eqs. (3.1) and (3.2) that the gain-floor v for the player I is closely related to all values a_{ij}, i.e., entries in the payoff matrix $\boldsymbol{A}$. In other words, v is a function of the values a_{ij} $(i = 1, 2, \ldots, m;\ j = 1, 2, \ldots, n)$ in the interval-valued payoffs $\bar{a}_{ij} = [a_{Lij}, a_{Rij}]$, denoted by $v = \upsilon((a_{ij}))$ or $v = \upsilon(\boldsymbol{A})$. Similarly, the optimal mixed strategy $\boldsymbol{y}^* \in Y$ of the player I in the constrained matrix game $\boldsymbol{A}$ is also a function of the values a_{ij} $(i = 1, 2, \ldots, m;\ j = 1, 2, \ldots, n)$, denoted by $\boldsymbol{y}^* = \boldsymbol{y}^*((a_{ij}))$ or $\boldsymbol{y}^* = \boldsymbol{y}^*(\boldsymbol{A})$.

According to Eqs. (3.1) and (3.2), it is easy to prove that the gain-floor $v = \upsilon((a_{ij}))$ for the player I is a non-decreasing function of the values a_{ij} $(i = 1, 2, \ldots, m;\ j = 1, 2, \ldots, n)$ in the interval-valued payoffs $\bar{a}_{ij} = [a_{Lij}, a_{Rij}]$. In fact, for any values a_{ij} and a'_{ij} in the interval-valued payoffs $\bar{a}_{ij} = [a_{Lij}, a_{Rij}]$ $(i = 1, 2, \ldots, m;\ j = 1, 2, \ldots, n)$, if $a_{ij} \le a'_{ij}$, then we have

$$\sum_{i=1}^{m}\sum_{j=1}^{n} y_i a_{ij} z_j \le \sum_{i=1}^{m}\sum_{j=1}^{n} y_i a'_{ij} z_j \tag{3.7}$$

since $y_i \ge 0$ $(i = 1, 2, \ldots, m)$ and $z_j \ge 0$ $(j = 1, 2, \ldots, n)$, where $\boldsymbol{y} \in Y$ and $\boldsymbol{z} \in Z$. Hence, we have

$$\min_{\boldsymbol{z}\in Z}\Big\{\sum_{i=1}^{m}\sum_{j=1}^{n} y_i a_{ij} z_j\Big\} \le \min_{\boldsymbol{z}\in Z}\Big\{\sum_{i=1}^{m}\sum_{j=1}^{n} y_i a'_{ij} z_j\Big\}, \tag{3.8}$$

which directly infers that

$$\max_{\boldsymbol{y}\in Y}\min_{\boldsymbol{z}\in Z}\Big\{\sum_{i=1}^{m}\sum_{j=1}^{n} y_i a_{ij} z_j\Big\} \le \max_{\boldsymbol{y}\in Y}\min_{\boldsymbol{z}\in Z}\Big\{\sum_{i=1}^{m}\sum_{j=1}^{n} y_i a'_{ij} z_j\Big\}, \tag{3.9}$$

i.e.,

$$\upsilon((a_{ij})) \leq \upsilon((a'_{ij})), \tag{3.10}$$

or $\upsilon(\boldsymbol{A}) \leq \upsilon(\boldsymbol{A}')$, where $\boldsymbol{A}' = (a'_{ij})_{m\times n}$ is the payoff matrix of the constrained matrix game $\boldsymbol{A}'$.

3.3.2 Linear Programming Methods of Interval-Valued Constrained Matrix Games

Because the expected payoff of the player I is a linear combination of interval-valued payoffs $\bar{a}_{ij} = [a_{Lij}, a_{Rij}]$, the gain-floor for the player I should be a closed interval as well. Stated as above, the gain-floor $v = \upsilon((a_{ij}))$ for the player I is a non-decreasing function of the values a_{ij} ($i = 1, 2, \ldots, m$; $j = 1, 2, \ldots, n$) in the interval-valued payoffs $\bar{a}_{ij} = [a_{Lij}, a_{Rij}]$. Hence, the upper bound v_R of the gain-floor $\bar{v}$ for the player I can be obtained as follows:

$$v_R = \max_{y\in Y}\min_{z\in Z}\{\boldsymbol{y}^{\mathrm{T}}\boldsymbol{A}_R\boldsymbol{z}\} = \max_{y\in Y}\min_{z\in Z}\{\sum_{i=1}^{m}\sum_{j=1}^{n} y_i a_{Rij} z_j\}, \tag{3.11}$$

where $\boldsymbol{A}_R = (a_{Rij})_{m\times n}$.

According to Eq. (3.5), Eq. (3.11) is equivalent to the linear programming model as follows:

$$\max\{\boldsymbol{d}^{\mathrm{T}}\boldsymbol{x}_R\}$$
$$\text{s.t.}\begin{cases} \boldsymbol{E}^{\mathrm{T}}\boldsymbol{x}_R - \boldsymbol{A}_R^{\mathrm{T}}\boldsymbol{y}_R \leq \boldsymbol{0} \\ \boldsymbol{B}^{\mathrm{T}}\boldsymbol{y}_R \leq \boldsymbol{c} \\ \boldsymbol{x}_R \geq \boldsymbol{0} \\ \boldsymbol{y}_R \geq \boldsymbol{0}, \end{cases} \tag{3.12}$$

where $\boldsymbol{x}_R$ and $\boldsymbol{y}_R$ are vectors of decision variables.

If Eq. (3.12) is feasible linear programming, then using the simplex method of linear programming, we can obtain its optimal solution, denoted by $(\boldsymbol{x}_R^*, \boldsymbol{y}_R^*)$. Thus, according to Theorem 3.2, we can obtain the upper bound $v_R = \boldsymbol{d}^{\mathrm{T}}\boldsymbol{x}_R^*$ of the player I's gain-floor $\bar{v}$ and corresponding optimal mixed strategy $\boldsymbol{y}_R^* \in Y$ for the interval-valued constrained matrix game $\bar{\boldsymbol{A}}$.

Analogously, the lower bound v_L of the player I's gain-floor $\bar{v}$ and optimal mixed strategy $\boldsymbol{y}_L^* \in Y$ for the interval-valued constrained matrix game $\bar{\boldsymbol{A}}$ are $v_L = \upsilon((a_{Lij}))$ and $\boldsymbol{y}_L^* = \boldsymbol{y}^*((a_{Lij}))$, respectively. According to Eq. (3.5), $(v_L, \boldsymbol{y}_L^*)$ can be obtained by solving the linear programming model as follows:

$$\max\{\boldsymbol{d}^{\mathrm{T}}\boldsymbol{x}_L\}$$
$$\text{s.t.}\begin{cases} \boldsymbol{E}^{\mathrm{T}}\boldsymbol{x}_L - \boldsymbol{A}_L^{\mathrm{T}}\boldsymbol{y}_L \leq 0 \\ \boldsymbol{B}^{\mathrm{T}}\boldsymbol{y}_L \leq \boldsymbol{c} \\ \boldsymbol{x}_L \geq 0 \\ \boldsymbol{y}_L \geq 0, \end{cases} \tag{3.13}$$

where $\boldsymbol{x}_L$ and $\boldsymbol{y}_L$ are vectors of decision variables, $\boldsymbol{A}_L=(a_{Lij})_{m\times n}$.

If Eq. (3.13) is feasible linear programming, then we can obtain its optimal solution, denoted by $(\boldsymbol{y}_L^*, \boldsymbol{x}_L^*)^{\mathrm{T}}$. Thus, according to Theorem 3.2, we obtain the lower bound $v_L = \boldsymbol{d}^T\boldsymbol{x}_L^*$ of the player I's gain-floor $\bar{v}$ and corresponding optimal mixed strategy $\boldsymbol{y}_L^*$ for the interval-valued constrained matrix game $\bar{\boldsymbol{A}}$.

In a similar way to the above analysis, the loss-ceiling μ and optimal mixed strategy $\boldsymbol{z}^* \in Z$ for the player II in the constrained matrix game $\boldsymbol{A}$ are functions of the values a_{ij} $(i = 1, 2, \ldots, m;\ j = 1, 2, \ldots, n)$ in the interval-valued payoffs $\bar{a}_{ij} = [a_{Lij}, a_{Rij}]$, denoted by $\mu = \omega((a_{ij}))$ (or $\mu = \omega(\boldsymbol{A})$) and $\boldsymbol{z}^* = \boldsymbol{z}^*((a_{ij}))$ (or $\boldsymbol{z}^* = \boldsymbol{z}^*(\boldsymbol{A})$), respectively. It is easy to prove that the loss-ceiling $\mu = \omega((a_{ij}))$ of the player II is also a non-decreasing function of the values a_{ij} $(i = 1, 2, \ldots, m;$ $j = 1, 2, \ldots, n)$ in the interval-valued payoffs $\bar{a}_{ij} = [a_{Lij}, a_{Rij}]$. Thus, the upper bound μ_R of the loss-ceiling $\bar{\mu}$ for the player II in the interval-valued constrained matrix game $\bar{\boldsymbol{A}}$ and corresponding optimal strategy $\boldsymbol{z}_R^* \in Z$ are $\mu_R = \omega((a_{Rij}))$ and $\boldsymbol{z}_R^* = \boldsymbol{z}^*\left((a_{Rij})\right)$, respectively.

According to Eq. (3.6), $(\mu_R, \boldsymbol{z}_R^*)$ can be obtained by solving the linear programming model as follows:

$$\min\{\boldsymbol{c}^{\mathrm{T}}\boldsymbol{s}_R\}$$
$$\text{s.t.}\begin{cases} \boldsymbol{B}\boldsymbol{s}_R - \boldsymbol{A}_R\boldsymbol{z}_R \geq 0 \\ \boldsymbol{E}\boldsymbol{z}_R \geq \boldsymbol{d} \\ \boldsymbol{s}_R \geq 0 \\ \boldsymbol{z}_R \geq 0, \end{cases} \tag{3.14}$$

where $\boldsymbol{s}_R$ and $\boldsymbol{z}_R$ are vectors of decision variables.

If Eq. (3.14) is feasible linear programming, then using the simplex method of linear programming, we can obtain its optimal solution, denoted by $(\boldsymbol{z}_R^*, \boldsymbol{s}_R^*)$. Thus, according to Theorem 3.2, we obtain the upper bound $\mu_R = \boldsymbol{c}^{\mathrm{T}}\boldsymbol{s}_R^*$ of the loss-ceiling $\bar{\mu}$ and corresponding optimal mixed strategy $\boldsymbol{z}_R^*$ for the player II.

Similarly, the lower bound μ_L of the player II's loss-ceiling $\bar{\mu}$ and corresponding optimal mixed strategy $\boldsymbol{z}_L^* \in Z$ are $\mu_L = \omega((a_{Lij}))$ and $\boldsymbol{z}_L^* = \boldsymbol{z}^*((a_{Lij}))$, respectively. According to Eq. (3.6), $(\mu_L, \boldsymbol{z}_L^*)$ can be obtained by solving the linear programming model as follows:

$$\min\{\boldsymbol{c}^{\mathrm{T}}\boldsymbol{s}_L\}$$
$$\text{s.t.}\begin{cases}\boldsymbol{B}\boldsymbol{s}_L - \boldsymbol{A}_L\boldsymbol{z}_L \geq 0\\ \boldsymbol{E}\boldsymbol{z}_L \geq \boldsymbol{d}\\ \boldsymbol{s}_L \geq 0\\ \boldsymbol{z}_L \geq 0,\end{cases} \tag{3.15}$$

where $\boldsymbol{s}_L$ and $\boldsymbol{z}_L$ are vectors of decision variables.

If Eq. (3.15) is feasible linear programming, then we can obtain its optimal solution, denoted by $(\boldsymbol{z}_L^*, \boldsymbol{s}_L^*)^{\mathrm{T}}$. Thus, we obtain the lower bound $\mu_L = \boldsymbol{c}^{\mathrm{T}}\boldsymbol{s}_L^*$ of the loss-ceiling $\bar{\mu}$ and corresponding optimal mixed strategy $\boldsymbol{z}_L^*$ for the player II in the interval-valued constrained matrix game $\bar{\boldsymbol{A}}$.

It is easy to see that Eqs. (3.13) and (3.15) are a pair of primal-dual linear programming. Then, by the duality theorem of linear programming, the lower bound v_L of the player I's gain-floor is equal to the lower bound μ_L of the player II's loss-ceiling, i.e., $v_L = \mu_L$. Likewise, Eqs. (3.12) and (3.14) are a pair of primal-dual linear programming. Hence, $v_R = \mu_R$. Therefore, the players I and II have the identical interval-type value, i.e., $[v_L, v_R] = [\mu_L, \mu_R]$. Hereby, the interval-valued constrained matrix game $\bar{\boldsymbol{A}}$ has a value $\bar{V}$, which is also a closed interval $\bar{V} = [V_L, V_R]$, where $V_R = v_R = \mu_R$ and $V_L = v_L = \mu_L$. Namely, $\bar{V} = \bar{v} = \bar{\mu}$.

3.3.3 Real Example Analysis of Market Share Problems

Let us consider the following numerical example of the market share game problem.

Suppose that there are two companies p_1 and p_2 aiming to enhance the market share of a product in a targeted market under the circumstance that the demand amount of the product in the targeted market basically is fixed. In other words, the market share of one company increases while the market share of another company decreases. The two companies are considering about two options (i.e., pure strategies) to increase the market share: improving technology (δ_1), advertisement (δ_2). The company p_1 needs the funds 80 and 50 (million Yuan) when it takes the pure strategies δ_1 and δ_2, respectively. However, due to the lack of the funds, the company p_1 only provides 60 (million Yuan), i.e., the mixed strategies of the company p_1 must satisfy the constraint condition: $80y_1 + 50y_2 \leq 60$. Similarly, the company p_2 needs the funds 40 and 70 (million Yuan) when it takes the pure strategies δ_1 and δ_2, respectively. However, the company p_2 only provides 50 (million Yuan), i.e., the mixed strategies of the company p_2 must satisfy the constraint condition: $40z_1 + 70z_2 \leq 50$ or $-40z_1 - 70z_2 \geq -50$. Due to a lack of information or imprecision of the available information, the managers of the two companies usually are not able to exactly forecast the sales amount of the companies' product. Hence, the interval can suitably represent the sales amount of the product from both companies' perspectives. Thus, the above problem may be regarded as an interval-valued constrained matrix game $\bar{\boldsymbol{A}}_0$. Namely, the companies p_1 and p_2 are regarded as the players I and II, respectively. The constrained sets of mixed strategies are expressed as follows:

$$Y_0 = \{\boldsymbol{y} | 80y_1 + 50y_2 \leq 60, y_1 + y_2 \leq 1, -y_1 - y_2 \leq -1, y_1 \geq 0, y_2 \geq 0\}$$

and

$$Z_0 = \{\boldsymbol{z} | -40z_1 - 70z_2 \geq -50, z_1 + z_2 \leq 1, -z_1 - z_2 \leq -1, z_1 \geq 0, z_2 \geq 0\},$$

respectively. The interval-valued payoff matrix $\bar{\boldsymbol{A}}_0$ of the player I (i.e., company p_1) is given as follows:

$$\bar{\boldsymbol{A}}_0 = \begin{matrix} & \begin{matrix} \delta_1 & \quad\quad\quad \delta_2 \end{matrix} \\ \begin{matrix} \delta_1 \\ \delta_2 \end{matrix} & \begin{pmatrix} [27, 35] & [-25, -17] \\ [-11, -5] & [35, 41] \end{pmatrix} \end{matrix},$$

where the element [27, 35] in $\bar{\boldsymbol{A}}_0$ is an interval, which indicates that the sales amount of the product for the company p_1 is between 27 and 35 when the companies p_1 and p_2 use the pure strategy δ_1 (improving technology) simultaneously. Other elements (i.e., intervals) in $\bar{\boldsymbol{A}}_0$ can be explained similarly.

Coefficient matrices and vectors of the constraint sets of mixed strategies for the companies p_1 and p_2 are obtained as follows:

$$\boldsymbol{B}_0 = \begin{pmatrix} 80 & 1 & -1 \\ 50 & 1 & -1 \end{pmatrix}, \quad \boldsymbol{E}_0^{\mathrm{T}} = \begin{pmatrix} -40 & 1 & -1 \\ -70 & 1 & -1 \end{pmatrix}$$

and

$$\boldsymbol{c}_0 = (60, 1, -1)^{\mathrm{T}}, \; \boldsymbol{d}_0 = (-50, 1, -1)^{\mathrm{T}},$$

respectively.

According to Eq. (3.12), the linear programming model can be constructed as follows:

$$\max\{-50x_{R1} + x_{R2} - x_{R3}\}$$
$$\text{s.t.} \begin{cases} -40x_{R1} + x_{R2} - x_{R3} - 35y_{R1} + 5y_{R2} \leq 0 \\ -70x_{R1} + x_{R2} - x_{R3} + 17y_{R1} - 41y_{R2} \leq 0 \\ 80y_{R1} + 50y_{R2} \leq 60 \\ y_{R1} + y_{R2} \leq 1 \\ -y_{R1} - y_{R2} \leq -1 \\ x_{R1} \geq 0, x_{R2} \geq 0, x_{R3} \geq 0, y_{R1} \geq 0, y_{R2} \geq 0, \end{cases} \tag{3.16}$$

where x_{R1}, x_{R2}, x_{R3}, y_{R1}, and y_{R2} are decision variables.

Solving Eq. (3.16) by using the simplex method of linear programming, we obtain its optimal solution $(\boldsymbol{x}_R^*, \boldsymbol{y}_R^*)$, where $\boldsymbol{y}_R^* = (1/3, 2/3)^{\mathrm{T}}$ and $\boldsymbol{x}_R^* = (0, 8.333, 0)^{\mathrm{T}}$. Therefore, the upper bound v_R of the gain-floor for the company p_1 and corresponding optimal mixed strategy $\boldsymbol{y}_R^*$ are $v_R = \boldsymbol{d}^{\mathrm{T}}\boldsymbol{x}_R^* = 8.333$ and $\boldsymbol{y}_R^* = (1/3, 2/3)^{\mathrm{T}}$, respectively.

Analogously, according to Eq. (3.13), the linear programming model can be obtained as follows:

$$\max\{-50x_{L1} + x_{L2} - x_{L3}\}$$
$$\text{s.t.}\begin{cases} -40x_{L1} + x_{L2} - x_{L3} - 27y_{L1} + 11y_{L2} \le 0 \\ -70x_{L1} + x_{L2} - x_{L3} + 25y_{L1} - 35y_{L2} \le 0 \\ 80y_{L1} + 50y_{L2} \le 60 \\ y_{L1} + y_{L2} \le 1 \\ -y_{L1} - y_{L2} \le -1 \\ x_{L1} \ge 0, x_{L2} \ge 0, x_{L3} \ge 0, y_{L1} \ge 0, y_{L2} \ge 0, \end{cases} \tag{3.17}$$

where x_{L1}, x_{L2}, x_{L3}, y_{L1}, and y_{L2} are decision variables.

Solving Eq. (3.17) by using the simplex method of linear programming, we can obtain its optimal solution $(\boldsymbol{x}_L^*, \boldsymbol{y}_L^*)$, where $\boldsymbol{y}_L^* = (1/3, 2/3)^{\mathrm{T}}$ and $\boldsymbol{x}_L^* = (0, 1.667, 0)^{\mathrm{T}}$. Therefore, the lower bound v_L of the gain-floor for the company p_1 and corresponding optimal mixed strategy $\boldsymbol{y}_L^*$ are $v_L = \boldsymbol{d}^{\mathrm{T}}\boldsymbol{x}_L^* = 1.667$ and $\boldsymbol{y}_L^* = (1/3, 2/3)^{\mathrm{T}}$, respectively. Thus, the gain-floor of the company p_1 is a closed interval $[v_L, v_R] = [1.667, 8.333]$, i.e., $\bar{v} = [1.667, 8.333]$.

In the same way, according to Eq. (3.14), the linear programming model can be obtained as follows:

$$\min\{60s_{R1} + s_{R2} - s_{R3}\}$$
$$\text{s.t.}\begin{cases} 80s_{R1} + s_{R2} - s_{R3} - 35z_{R1} + 17z_{R2} \ge 0 \\ 50s_{R1} + s_{R2} - s_{R3} + 5z_{R1} - 41z_{R2} \ge 0 \\ -40z_{R1} - 70z_{R2} \ge -50 \\ z_{R1} + z_{R2} \ge 1 \\ -z_{R1} - z_{R2} \ge -1 \\ s_{R1} \ge 0, s_{R2} \ge 0, s_{R3} \ge 0, z_{R1} \ge 0, z_{R2} \ge 0, \end{cases} \tag{3.18}$$

where s_{R1}, s_{R2}, s_{R3}, z_{R1}, and z_{R2} are decision variables.

Solving Eq. (3.18) by using the simplex method of linear programming, we obtain its optimal solution $(\boldsymbol{s}_R^*, \boldsymbol{z}_R^*)$, where $\boldsymbol{z}_R^* = (1, 0)^{\mathrm{T}}$ and $\boldsymbol{s}_R^* = (1.333, 0, 71.667)^{\mathrm{T}}$, respectively. Therefore, the upper bound μ_R of the loss-ceiling for the company p_2 and corresponding optimal mixed strategy z_R^* are $\mu_R = \boldsymbol{d}^{\mathrm{T}} \boldsymbol{s}_R^* = 8.333$ and $\boldsymbol{z}_R^* = (1, 0)^{\mathrm{T}}$, respectively.

According to Eq. (3.15), the linear programming model can be obtained as follows:

$$\min\{60 s_{L1} + s_{L2} - s_{L3}\}$$
$$\text{s.t.} \begin{cases} 8 s_{L1} + s_{L2} - s_{L3} - 27 z_{L1} + 25 z_{L2} \geq 0 \\ 50 s_{L1} + s_{L2} - s_{L3} + 11 z_{L1} - 35 z_{L2} \geq 0 \\ -40 z_{L1} - 70 z_{L2} \geq -50 \\ z_{L1} + z_{L2} \geq 1 \\ -z_{L1} - z_{L2} \geq -1 \\ s_{L1} \geq 0, s_{L2} \geq 0, s_{L3} \geq 0, z_{L1} \geq 0, z_{L2} \geq 0, \end{cases} \quad (3.19)$$

where s_{L1}, s_{L2}, s_{L3}, z_{L1}, and z_{L2} are decision variables.

Solving Eq. (3.19) by using the simplex method of linear programming, we obtain its optimal solution $(\boldsymbol{z}_L^*, \boldsymbol{s}_L^*)$, where $\boldsymbol{z}_L^* = (1, 0)^{\mathrm{T}}$ and $\boldsymbol{s}_L^* = (1.267, 0, 74.333)^{\mathrm{T}}$. Therefore, the lower bound μ_L of the loss-ceiling for the company p_2 and corresponding optimal mixed strategy $\boldsymbol{z}_L^*$ are $\mu_L = \boldsymbol{d}^{\mathrm{T}} \boldsymbol{s}_L^* = 1.667$ and $\boldsymbol{z}_L^* = (1, 0)^{\mathrm{T}}$, respectively. Thus, the loss-ceiling of the company p_2 is a closed interval $[\mu_L, \mu_R] = [1.667, 8.333]$, i.e., $\bar{\mu} = [1.667, 8.333]$.

Obviously, $\bar{v} = \bar{\mu} = [1.667, 8.333]$, i.e., the companies p_1 and p_2 have the identical interval-type value. Therefore, the interval-valued constrained matrix game $\bar{\boldsymbol{A}}_0$ has a value, which is an interval $\bar{V} = \bar{v} = \bar{\mu} = [1.667, 8.333]$.

Conversely, if both companies do not take into account the constraints of the strategies, then the above market share problem may be regarded as an interval-valued (unconstraint) matrix game $\bar{\boldsymbol{A}}_0$ [9]. Thus, according to Eqs. (3.13) and (3.19) given by Li [9], we have

$$v'_R = 13.776, \; \boldsymbol{y}'^*_R = (0.469, 0.531)^{\mathrm{T}}$$

and

$$v'_L = 6.837, \; \boldsymbol{y}'^*_L = (0.469, 0.531)^{\mathrm{T}}.$$

Therefore, the gain-floor of the company p_1 is a closed interval $\bar{v}' = [v'_L, v'_R] = [6.837, 13.776]$.

Likewise, according to Eqs. (3.25) and (3.31) given by Li [9], we have

$$\mu'_R = 13.776, \; \boldsymbol{z}_R^* = (0.592, 0.408)^{\mathrm{T}}$$

and

$$\mu'_L = 6.837,\ z'_L = (0.612, 0.388).$$

Then, the loss-ceiling of the company p_2 is a closed interval $\bar{\mu}' = [\mu'_L, \mu'_R] = [6.837, 13.776]$.

Thus, the companies p_1 and p_2 have the identical interval-type value. Hereby, the interval-valued (unconstraint) matrix game $\bar{A}_0$ has a value $\bar{V}' = \bar{v}' = \bar{\mu}' = [6.837, 13.776]$.

Obviously, the value $\bar{V}' = [6.837, 13.776]$ and corresponding optimal mixed strategies for the companies p_1 and p_2 in the interval-valued unconstrained matrix game are different from the value $\bar{V} = [1.667, 8.333]$ and corresponding optimal mixed strategies in the interval-valued constrained matrix game. Moreover, $\bar{V}' = [6.837, 13.776]$ is larger than $\bar{V} = [1.667, 8.333]$ since $\mu'_L > \mu_L$ and $\mu'_R > \mu_R$ according to the order relations over intervals [8]. This conclusion is accordance with the actual situation as expected. On the other hand, it is shown that it is necessary to consider the constraint conditions of strategies.

References

1. Owen G (1982) Game theory, 2nd edn. Academic Press, New York
2. Nishizaki I, Sakawa M (2001) Fuzzy and multiobjective games for conflict resolution. Physica-Verlag, Springer, Berlin
3. Li D-F (2003) Fuzzy multiobjective many-person decision makings and games. National Defense Industry Press, Beijing (in Chinese)
4. Bector CR, Chandra S (2005) Fuzzy mathematical programming and fuzzy matrix games. Springer, Berlin
5. Dresher M (1961) Games of strategy theory and applications. Prentice-Hall, New York
6. Li D-F (1999) Fuzzy constrained matrix games with fuzzy payoffs. J Fuzzy Math 7(4):873–880
7. Li D-F, Cheng C-T (2002) Fuzzy multiobjective programming methods for fuzzy constrained matrix games with fuzzy numbers. Int J Uncertainty, Fuzziness and Knowledge-Based Syst 10 (4):385–400
8. Moore RE (1979) Method and application of interval analysis. SIAM, Philadelphia
9. Li D-F (2011) Linear programming approach to solve interval-valued matrix games. Omega: Int J Manag Sci 39(6):655–666

Chapter 4
Constrained Matrix Games with Payoffs of Triangular Fuzzy Numbers

4.1 Introduction

As stated in previous three chapters, in real competitive or antagonistic situations, players cannot exactly estimate their payoffs due to lack of adequate information and/or imprecision of the available information on the environments. Thus, in Chap. 3, intervals are used to deal with imprecision of payoffs and hereby we have studied interval-valued constrained matrix games. In a parallel way to Chap. 2, triangular fuzzy numbers are used to appropriately model imprecision and/or vagueness of players' payoffs [1]. In fact, triangular fuzzy numbers may be regarded as a generalization of intervals. Due to the fact that choice of strategies for players is constrained in some real situations, it is necessary to study a type of constrained matrix games with payoffs expressed by triangular fuzzy numbers, which often are called constrained matrix games with payoffs of triangular fuzzy numbers for short [2, 3]. Obviously, a constrained matrix game with payoffs of triangular fuzzy numbers is remarkably different from a classical matrix game or fuzzy matrix game as in Chaps. 1 and 2 in that the former simultaneously involves the payoffs' fuzziness and strategies' constraints. Any classical matrix game always has a value in the sense of mixed strategies and optimal mixed strategies of players, which are desirable and can be easily obtained by solving a pair of primal-dual linear programming models [4, 5]. However, there is no effective and efficient method which can always ensure that any fuzzy matrix game has a fuzzy value. As a result, the methods of classical and fuzzy matrix games are not applicable to constrained matrix games with payoffs of triangular fuzzy numbers. As far as we know, in fact, there is no effective and efficient method for solving constrained matrix games with payoffs of triangular fuzzy numbers. In this chapter, we focus on developing the fuzzy multi-objective programming method and linear programming method for solving constrained matrix games with payoffs of triangular fuzzy numbers.

D.-F. Li, *Linear Programming Models and Methods of Matrix Games with Payoffs of Triangular Fuzzy Numbers*, Studies in Fuzziness and Soft Computing 328 ,
DOI 10.1007/978-3-662-48476-0_4

4.2 Fuzzy Multi-Objective Programming Models of Constrained Matrix Games with Payoffs of Triangular Fuzzy Numbers

4.2.1 *Constrained Matrix Games with Payoffs of Triangular Fuzzy Numbers*

As data which define the restrictions of the sets of possible (mixed) strategies of the players I and II are fuzzy, some violation in the accomplishment of the constraints should be allowed. Consequently, the strategic possibilities of the players would be defined in a vague way, i.e., $\tilde{Y}$ and $\tilde{Z}$ would be fuzzy sets on the sets S_1 and S_2 of pure strategies, which are defined as in Sect. 1.2. Qualitatively, a constrained matrix game with fuzzy data is called a fuzzy constrained matrix game.

Mathematically, without loss of generality, a fuzzy payoff matrix of the player I is defined as $\tilde{\boldsymbol{A}} = (\tilde{a}_{ij})_{m\times n}$, which is defined as in Sect. 2.3.2. More precisely,

$$\tilde{\boldsymbol{A}} = (\tilde{a}_{ij})_{m\times n} = \begin{matrix} \\ \delta_1 \\ \delta_2 \\ \vdots \\ \delta_m \end{matrix} \begin{matrix} \begin{matrix} \beta_1 & \beta_2 & \cdots & \beta_n \end{matrix} \\ \begin{pmatrix} \tilde{a}_{11} & \tilde{a}_{12} & \cdots & \tilde{a}_{1n} \\ \tilde{a}_{21} & \tilde{a}_{22} & \cdots & \tilde{a}_{2n} \\ \vdots & \vdots & \cdots & \vdots \\ \tilde{a}_{m1} & \tilde{a}_{m2} & \cdots & \tilde{a}_{mn} \end{pmatrix} \end{matrix},$$

where $\tilde{a}_{ij} = (a^l_{ij}, a^m_{ij}, a^r_{ij})$ $(i = 1, 2, \ldots, m; j = 1, 2, \ldots, n)$ are triangular fuzzy numbers defined as in Sect. 2.2. $\tilde{Y} = \{\boldsymbol{y} | \tilde{\boldsymbol{B}}^{\mathrm{T}} \boldsymbol{y} \leq \tilde{\boldsymbol{c}}, \boldsymbol{y} \geq \boldsymbol{0}\}$ and $\tilde{Z} = \{\boldsymbol{z} | \tilde{\boldsymbol{E}} \boldsymbol{z} \geq \tilde{\boldsymbol{d}}, \boldsymbol{z} \geq \boldsymbol{0}\}$ represent the fuzzy constraint sets of strategies for players I and II, where $\tilde{\boldsymbol{c}} = (\tilde{c}_1, \tilde{c}_2, \ldots, \tilde{c}_p)^{\mathrm{T}}$ and $\tilde{\boldsymbol{d}} = (\tilde{d}_1, \tilde{d}_2, \ldots, \tilde{d}_q)^{\mathrm{T}}$ are vectors of triangular fuzzy numbers, and $\tilde{\boldsymbol{B}} = (\tilde{b}_{il})_{m\times p}$ and $\tilde{\boldsymbol{E}} = (\tilde{e}_{kj})_{q\times n}$ are matrixes of triangular fuzzy numbers, $\tilde{c}_h = (c^l_h, c^m_h, c^r_h)$ $(h = 1, 2, \ldots, p)$, $\tilde{d}_k = (d^l_k, d^m_k, d^r_k)$ $(k = 1, 2, \ldots, q)$, $\tilde{b}_{ih} = (b^l_{ih}, b^m_{ih}, b^r_{ih})$ $(i = 1, 2, \ldots, m; h = 1, 2, \ldots, p)$, and $\tilde{e}_{kj} = (e^l_{kj}, e^m_{kj}, e^r_{kj})$ $(k = 1, 2, \ldots, q; j = 1, 2, \ldots, n)$ are triangular fuzzy numbers. Thus, a constrained matrix game with payoffs of triangular fuzzy numbers is meant that the payoff matrix of the player I is $\tilde{\boldsymbol{A}}$ (hereby the payoff matrix of the player II is $-\tilde{\boldsymbol{A}}$) and the constraint sets of strategies for the players I and II are $\tilde{Y}$ and $\tilde{Z}$, respectively. In the sequent, it is often called the constrained matrix game $\tilde{\boldsymbol{A}}$ with payoffs of triangular fuzzy numbers.

Thus, in a parallel way to the auxiliary linear programming models [i.e., Eqs. (3.5) and (3.6)], we can construct the fuzzy mathematical programming models as follows:

$$\max\{\tilde{\boldsymbol{d}}^{\mathrm{T}}\boldsymbol{x}\}$$
$$\text{s.t.}\begin{cases}\tilde{\boldsymbol{E}}^{\mathrm{T}}\boldsymbol{x}\tilde{\leq}\tilde{\boldsymbol{A}}^{\mathrm{T}}\boldsymbol{y}\\ \tilde{\boldsymbol{B}}^{\mathrm{T}}\boldsymbol{y}\tilde{\leq}\tilde{\boldsymbol{c}}\\ \boldsymbol{x}\geq\boldsymbol{0}\\ \boldsymbol{y}\geq\boldsymbol{0}\end{cases}\tag{4.1}$$

and

$$\min\{\tilde{\boldsymbol{c}}^{\mathrm{T}}\boldsymbol{s}\}$$
$$\text{s.t.}\begin{cases}\tilde{\boldsymbol{B}}\boldsymbol{s}\tilde{\geq}\tilde{\boldsymbol{A}}\boldsymbol{z}\\ \tilde{\boldsymbol{E}}\boldsymbol{z}\tilde{\geq}\tilde{\boldsymbol{d}}\\ \boldsymbol{s}\geq\boldsymbol{0}\\ \boldsymbol{z}\geq\boldsymbol{0},\end{cases}\tag{4.2}$$

respectively, where $\boldsymbol{y}$, $\boldsymbol{x}$, $\boldsymbol{s}$, and $\boldsymbol{z}$ are vectors of decision variables.

If $(\boldsymbol{y}^*,\boldsymbol{x}^*)^{\mathrm{T}}$ is an optimal solution of Eq. (4.1), $\boldsymbol{y}^*$ is called a maximin (mixed) strategy of the player I in the constrained matrix game $\tilde{\boldsymbol{A}}$ with payoffs of triangular fuzzy numbers. Usually, $\boldsymbol{y}^*$ is briefly called an optimal (mixed) strategy of the player I. Similarly, if $(\boldsymbol{z}^*,\boldsymbol{s}^*)^{\mathrm{T}}$ is an optimal solution of Eq. (4.2), $\boldsymbol{z}^*$ is called a minimax (mixed) strategy of the player II in the constrained matrix game $\tilde{\boldsymbol{A}}$ with payoffs of triangular fuzzy numbers. Briefly, $\boldsymbol{z}^*$ is called an optimal (mixed) strategy of the player II. If $\boldsymbol{y}^*$ is a maximin (mixed) strategy of the player I and $\boldsymbol{z}^*$ is a minimax (mixed) strategy of the player II, $(\boldsymbol{y}^*,\boldsymbol{z}^*)^{\mathrm{T}}$ is called a solution of the constrained matrix game $\tilde{\boldsymbol{A}}$ with payoffs of triangular fuzzy numbers. Denote

$$\tilde{\upsilon}^* = \tilde{\boldsymbol{d}}^{\mathrm{T}}\boldsymbol{x}^*$$

and

$$\tilde{\omega}^* = \tilde{\boldsymbol{c}}^{\mathrm{T}}\boldsymbol{s}^*.$$

Then, $\tilde{\upsilon}^*$ and $\tilde{\omega}^*$ are called the player I's gain-floor and the player II's loss-ceiling, respectively. Let

$$\tilde{V}^* = \tilde{\upsilon}^* \wedge \tilde{\omega}^*.\tag{4.3}$$

Then, $\tilde{V}^*$ is called a fuzzy equilibrium value of the constrained matrix game $\tilde{\boldsymbol{A}}$ with payoffs of triangular fuzzy numbers.

Using the operations of triangular fuzzy numbers [i.e., Eqs. (2.2) and (2.3)] and their order relations (i.e., Definition 2.1, with reference to [6]), in general we draw the following conclusion, which is summarized as in Theorem 4.1.

Theorem 4.1 *Assume that* $(\boldsymbol{y}^*,\boldsymbol{x}^*)^{\mathrm{T}}$ *and* $(\boldsymbol{z}^*,\boldsymbol{s}^*)^{\mathrm{T}}$ *are optimal solutions of* Eqs. (4.1) and (4.2), respectively. Then, $\tilde{\upsilon}^*$ *and* $\tilde{\omega}^*$ *are triangular fuzzy numbers and* $\tilde{\upsilon}^* \tilde{\le} \tilde{\omega}^*$.

Proof According to Eqs. (2.2) and (2.3), it is obviously that $\tilde{\upsilon}^*$ and $\tilde{\omega}^*$ are triangular fuzzy numbers.

Due to the assumption that $(\boldsymbol{y}^*,\boldsymbol{x}^*)^{\mathrm{T}}$ and $(\boldsymbol{z}^*,\boldsymbol{s}^*)^{\mathrm{T}}$ are optimal solutions of Eqs. (4.1) and (4.2), respectively, then we have

$$\begin{aligned}
&\tilde{\boldsymbol{B}}^{\mathrm{T}}\boldsymbol{y}^* \tilde{\le} \tilde{\boldsymbol{c}},\\
&\tilde{\boldsymbol{E}}^{\mathrm{T}}\boldsymbol{x}^* \tilde{\le} \tilde{\boldsymbol{A}}^{\mathrm{T}}\boldsymbol{y}^*,\\
&\tilde{\boldsymbol{A}}\boldsymbol{z}^* \tilde{\le} \tilde{\boldsymbol{B}}\boldsymbol{s}^*,\\
&\tilde{\boldsymbol{d}} \tilde{\le} \tilde{\boldsymbol{E}}\boldsymbol{z}^*
\end{aligned}$$

and

$$\boldsymbol{x}^* \ge 0,\ \boldsymbol{y}^* \ge 0,\ \boldsymbol{z}^* \ge 0,\ \boldsymbol{s}^* \ge 0.$$

Hereby, combining with Eqs. (2.2) and (2.3) and Definition 2.1, we obtain

$$\begin{aligned}
\tilde{\boldsymbol{d}}^{\mathrm{T}}\boldsymbol{x}^* \tilde{\le} (\tilde{\boldsymbol{E}}\boldsymbol{z}^*)^{\mathrm{T}}\boldsymbol{x}^* &= \boldsymbol{z}^{*\mathrm{T}}(\tilde{\boldsymbol{E}}^{\mathrm{T}}\boldsymbol{x}^*) \tilde{\le} \boldsymbol{z}^{*\mathrm{T}}(\tilde{\boldsymbol{A}}^{\mathrm{T}}\boldsymbol{y}^*)\\
&= (\tilde{\boldsymbol{A}}\boldsymbol{z}^*)^{\mathrm{T}}\boldsymbol{y}^* \tilde{\le} (\tilde{\boldsymbol{B}}\boldsymbol{s}^*)^{\mathrm{T}}\boldsymbol{y}^*\\
&= \boldsymbol{s}^{*\mathrm{T}}(\tilde{\boldsymbol{B}}^{\mathrm{T}}\boldsymbol{y}^*) \tilde{\le} \boldsymbol{s}^{*\mathrm{T}}\tilde{\boldsymbol{c}} = \tilde{\boldsymbol{c}}^{\mathrm{T}}\boldsymbol{s}^*,
\end{aligned}$$

i.e., $\tilde{\upsilon}^* \tilde{\le} \tilde{\omega}^*$. Thus, we have completed the proof of Theorem 4.1.

Theorem 4.1 means that the player I's gain-floor cannot exceed the player II's loss-ceiling, which is very similar to that of Theorem 2.2.

4.2.2 Fuzzy Multi-Objective Programming Method of Constrained Matrix Games with Payoffs of Triangular Fuzzy Numbers

In this subsection, we mainly study how to solve Eqs. (4.1) and (4.2) effectively. Namely, we focus on developing an effective and efficient method for solving the constrained matrix game $\tilde{\boldsymbol{A}}$ with payoffs of triangular fuzzy numbers.

Firstly, let us consider Eq. (4.1). According to Definition 2.1, Eq. (4.1) can be rewritten as the multi-objective mathematical programming model as follows:

$$\begin{aligned} &\max\{\boldsymbol{d}_l^{\mathrm{T}}\boldsymbol{x}\}\\ &\max\{\boldsymbol{d}_m^{\mathrm{T}}\boldsymbol{x}\}\\ &\max\{\boldsymbol{d}_r^{\mathrm{T}}\boldsymbol{x}\}\\ &\text{s.t.}\begin{cases} \tilde{\boldsymbol{E}}^{\mathrm{T}}\boldsymbol{x} \tilde{\leq} \tilde{\boldsymbol{A}}^{\mathrm{T}}\boldsymbol{y}\\ \tilde{\boldsymbol{B}}^{\mathrm{T}}\boldsymbol{y} \tilde{\leq} \tilde{\boldsymbol{c}}\\ \boldsymbol{x} \geq \boldsymbol{0}\\ \boldsymbol{y} \geq \boldsymbol{0}, \end{cases} \end{aligned} \tag{4.4}$$

where $\boldsymbol{d}_l = (d_1^l, d_2^l, \ldots, d_q^l)^{\mathrm{T}}$, $\boldsymbol{d}_m = (d_1^m, d_2^m, \ldots, d_q^m)^{\mathrm{T}}$, and $\boldsymbol{d}_r = (d_1^r, d_2^r, \ldots, d_q^r)^{\mathrm{T}}$.

Analogously, Eq. (4.2) can be rewritten as the multi-objective mathematical programming model as follows:

$$\begin{aligned} &\min\{\boldsymbol{c}_l^{\mathrm{T}}\boldsymbol{s}\}\\ &\min\{\boldsymbol{c}_m^{\mathrm{T}}\boldsymbol{s}\}\\ &\min\{\boldsymbol{c}_r^{\mathrm{T}}\boldsymbol{s}\}\\ &\text{s.t.}\begin{cases} \tilde{\boldsymbol{B}}\boldsymbol{s} \tilde{\geq} \tilde{\boldsymbol{A}}\boldsymbol{z}\\ \tilde{\boldsymbol{E}}\boldsymbol{z} \tilde{\geq} \tilde{\boldsymbol{d}}\\ \boldsymbol{s} \geq \boldsymbol{0}\\ \boldsymbol{z} \geq \boldsymbol{0}, \end{cases} \end{aligned} \tag{4.5}$$

where $\boldsymbol{c}_l = (c_1^l, c_2^l, \ldots, c_p^l)^{\mathrm{T}}$, $\boldsymbol{c}_m = (c_1^m, c_2^m, \ldots, c_p^m)^{\mathrm{T}}$, and $\boldsymbol{c}_r = (c_1^r, c_2^r, \ldots, c_p^r)^{\mathrm{T}}$.

Equations (4.4) and (4.5) are multi-objective programming. As stated earlier, they may be solved by using some multi-objective programming techniques such as utility theory, goal programming, fuzzy programming, and interactive approaches. In this subsection, we develop a fuzzy multi-objective programming method through using Zimmermann's fuzzy programming method [7] and auxiliary crisp inequality constraints with our normalization process.

It is obvious from Eqs. (4.4) and (4.5) that their constraint sets are fuzzy. Thus, our important problem is how to obtain auxiliary crisp constraint sets from the fuzzy constraint sets of Eqs. (4.4) and (4.5). One of possible resolutions for these fuzzy constraint sets is to use the concept of α-cuts of triangular fuzzy numbers and the weighted average method. Thus, we can obtain auxiliary crisp constraints as follows.

As stated in Sect. 2.2, any α-cut set of a triangular fuzzy number is a closed and bounded interval, where $\alpha \in [0,1]$. Thus, if the minimal acceptable possibility $\alpha \in [0,1]$ is given, then we can easily obtain α-cut sets of the triangular fuzzy numbers $\tilde{a}_{ij}$ and $\tilde{e}_{ij}$, i.e., $\tilde{a}_{ij}(\alpha) = [a_{ij}^L(\alpha), a_{ij}^R(\alpha)]$ and $\tilde{e}_{ij}(\alpha) = [e_{ij}^L(\alpha), e_{ij}^R(\alpha)]$. Denote $\boldsymbol{A}_L(\alpha) = (a_{ij}^L(\alpha))_{m\times n}$, $\boldsymbol{A}_m = (a_{ij}^m)_{m\times n}$, $\boldsymbol{A}_R(\alpha) = (a_{ij}^R(\alpha))_{m\times n}$, $\boldsymbol{E}_L(\alpha) = (e_{ij}^L(\alpha))_{q\times n}$, $\boldsymbol{E}_m = (e_{ij}^m)_{q\times n}$, and $\boldsymbol{E}_R(\alpha) = (e_{ij}^R(\alpha))_{q\times n}$. Then, by using the weighted average method, the fuzzy inequality constraint $\tilde{\boldsymbol{E}}^{\mathrm{T}}\boldsymbol{x} \tilde{\leq} \tilde{\boldsymbol{A}}^{\mathrm{T}}\boldsymbol{y}$ is converted into the auxiliary crisp inequality constraint as follows:

$$(\omega_1\boldsymbol{E}_L(\alpha) + \omega_2\boldsymbol{E}_m + \omega_3\boldsymbol{E}_R(\alpha))^{\mathrm{T}}\boldsymbol{x} \leq (\omega_1\boldsymbol{A}_L(\alpha) + \omega_2\boldsymbol{A}_m + \omega_3\boldsymbol{A}_R(\alpha))^{\mathrm{T}}\boldsymbol{y},$$

where $\omega_i \geq 0$ $(i = 1, 2, 3)$ are weights which satisfy the normalization condition: $\omega_1 + \omega_2 + \omega_3 = 1$. Weight determination methods may be referred to [8].

In the same way, the fuzzy inequalities constraints $\tilde{\boldsymbol{B}}^{\mathrm{T}}\boldsymbol{y} \tilde{\leq} \tilde{\boldsymbol{c}}$, $\tilde{\boldsymbol{B}}\boldsymbol{s} \tilde{\geq} \tilde{\boldsymbol{A}}\boldsymbol{z}$, and $\tilde{\boldsymbol{E}}\boldsymbol{z} \tilde{\geq} \tilde{\boldsymbol{d}}$ are converted into the auxiliary crisp inequality constraints as follows:

$$(\omega_1\boldsymbol{B}_L(\alpha) + \omega_2\boldsymbol{B}_m + \omega_3\boldsymbol{B}_R(\alpha))^{\mathrm{T}}\boldsymbol{y} \leq \omega_1\boldsymbol{c}_L(\alpha) + \omega_2\boldsymbol{c}_m + \omega_3\boldsymbol{c}_R(\alpha),$$
$$(\omega_1\boldsymbol{B}_L(\alpha) + \omega_2\boldsymbol{B}_m + \omega_3\boldsymbol{B}_R(\alpha))\boldsymbol{s} \geq (\omega_1\boldsymbol{A}_L(\alpha) + \omega_2\boldsymbol{A}_m + \omega_3\boldsymbol{A}_R(\alpha))\boldsymbol{z}$$

and

$$(\omega_1\boldsymbol{E}_L(\alpha) + \omega_2\boldsymbol{E}_m + \omega_3\boldsymbol{E}_R(\alpha))\boldsymbol{z} \geq \omega_1\boldsymbol{d}_L(\alpha) + \omega_2\boldsymbol{d}_m + \omega_3\boldsymbol{d}_R(\alpha),$$

respectively, where

$\tilde{b}_{ij}(\alpha) = [b_{ij}^L(\alpha), b_{ij}^R(\alpha)]$, $\tilde{c}_h(\alpha) = [c_h^L(\alpha), c_h^R(\alpha)]$, $\tilde{d}_k(\alpha) = [d_k^L(\alpha), d_k^R(\alpha)]$,
$\boldsymbol{B}_L(\alpha) = (b_{ij}^L(\alpha))_{m\times p}$, $\boldsymbol{B}_m = (b_{ij}^m)_{m\times p}$, $\boldsymbol{B}_R(\alpha) = (b_{ij}^R(\alpha))_{m\times p}$,
$\boldsymbol{c}_L(\alpha) = (c_1^L(\alpha), c_2^L(\alpha), \ldots, c_p^L(\alpha))^{\mathrm{T}}$, $\boldsymbol{c}_m = (c_1^m, c_2^m, \ldots, c_p^m)^{\mathrm{T}}$, $\boldsymbol{c}_R(\alpha) = (c_1^R(\alpha), c_2^R(\alpha), \ldots, c_p^R(\alpha))^{\mathrm{T}}$,
$\boldsymbol{d}_L(\alpha) = (d_1^L(\alpha), d_2^L(\alpha), \ldots, d_q^L(\alpha))^{\mathrm{T}}$, $\boldsymbol{d}_m = (d_1^m, d_2^m, \ldots, d_q^m)^{\mathrm{T}}$, and $\boldsymbol{d}_R(\alpha) = (d_1^R(\alpha), d_2^R(\alpha), \ldots, d_q^R(\alpha))^{\mathrm{T}}$.

With the assumption of $\omega_1 = \omega_3 = 1/6$ and $\omega_2 = 4/6$, the above auxiliary crisp inequality constraint

$$(\omega_1\boldsymbol{E}_L(\alpha) + \omega_2\boldsymbol{E}_m + \omega_3\boldsymbol{E}_R(\alpha))^{\mathrm{T}}\boldsymbol{x} \leq (\omega_1\boldsymbol{A}_L(\alpha) + \omega_2\boldsymbol{A}_m + \omega_3\boldsymbol{A}_R(\alpha))^{\mathrm{T}}\boldsymbol{y}$$

can be written as follows:

$$(\boldsymbol{E}_L(\alpha) + 4\boldsymbol{E}_m + \boldsymbol{E}_R(\alpha))^{\mathrm{T}}\boldsymbol{x} \leq (\boldsymbol{A}_L(\alpha) + 4\boldsymbol{A}_m + \boldsymbol{A}_R(\alpha))^{\mathrm{T}}\boldsymbol{y}.$$

The weights ω_1, ω_2, and ω_3 can be changed subjectively. The reason of using the above weighted average values is that $\boldsymbol{E}_L(\alpha)$ and $\boldsymbol{A}_L(\alpha)$ are too pessimistic whereas $\boldsymbol{E}_R(\alpha)$ and $\boldsymbol{A}_R(\alpha)$ are too optimistic. Of course, these boundary values may provide us boundary solutions. Besides, the means (or most possible values) $\boldsymbol{E}_m$ and $\boldsymbol{A}_m$ are often the most important ones. Thus, more weights should be assigned [9].

Analogously, the above auxiliary crisp inequality constraints

$$\begin{aligned}
&(\omega_1\boldsymbol{B}_L(\alpha) + \omega_2\boldsymbol{B}_m + \omega_3\boldsymbol{B}_R(\alpha))^{\mathrm{T}}\boldsymbol{y} \leq \omega_1\boldsymbol{c}_L(\alpha) + \omega_2\boldsymbol{c}_m + \omega_3\boldsymbol{c}_R(\alpha),\\
&(\omega_1\boldsymbol{B}_L(\alpha) + \omega_2\boldsymbol{B}_m + \omega_3\boldsymbol{B}_R(\alpha))\boldsymbol{s} \geq (\omega_1\boldsymbol{A}_L(\alpha) + \omega_2\boldsymbol{A}_m + \omega_3\boldsymbol{A}_R(\alpha))\boldsymbol{z}
\end{aligned}$$

and

$$(\omega_1\boldsymbol{E}_L(\alpha) + \omega_2\boldsymbol{E}_m + \omega_3\boldsymbol{E}_R(\alpha))\boldsymbol{z} \geq \omega_1\boldsymbol{d}_L(\alpha) + \omega_2\boldsymbol{d}_m + \omega_3\boldsymbol{d}_R(\alpha)$$

can be written as follows:

$$\begin{aligned}
&(\boldsymbol{B}_L(\alpha) + 4\boldsymbol{B}_m + \boldsymbol{B}_R(\alpha))^{\mathrm{T}}\boldsymbol{y} \leq \boldsymbol{c}_L(\alpha) + 4\boldsymbol{c}_m + \boldsymbol{c}_R(\alpha),\\
&(\boldsymbol{B}_L(\alpha) + 4\boldsymbol{B}_m + \boldsymbol{B}_R(\alpha))\boldsymbol{s} \geq (\boldsymbol{A}_L(\alpha) + 4\boldsymbol{A}_m + \boldsymbol{A}_R(\alpha))\boldsymbol{z}
\end{aligned}$$

and

$$(\boldsymbol{E}_L(\alpha) + 4\boldsymbol{E}_m + \boldsymbol{E}_R(\alpha))\boldsymbol{z} \geq \boldsymbol{d}_L(\alpha) + 4\boldsymbol{d}_m + \boldsymbol{d}_R(\alpha),$$

respectively.

Therefore, Eqs. (4.4) and (4.5) can be transformed into the multi-objective programming models as follows:

$$\begin{aligned}
&\max\{\boldsymbol{d}_l^{\mathrm{T}}\boldsymbol{x}\}\\
&\max\{\boldsymbol{d}_m^{\mathrm{T}}\boldsymbol{x}\}\\
&\max\{\boldsymbol{d}_r^{\mathrm{T}}\boldsymbol{x}\}\\
&\text{s.t.}\begin{cases}
(\boldsymbol{E}_L(\alpha) + 4\boldsymbol{E}_m + \boldsymbol{E}_R(\alpha))^{\mathrm{T}}\boldsymbol{x} \leq (\boldsymbol{A}_L(\alpha) + 4\boldsymbol{A}_m + \boldsymbol{A}_R(\alpha))^{\mathrm{T}}\boldsymbol{y}\\
(\boldsymbol{B}_L(\alpha) + 4\boldsymbol{B}_m + \boldsymbol{B}_R(\alpha))^{\mathrm{T}}\boldsymbol{y} \leq \boldsymbol{c}_L(\alpha) + 4\boldsymbol{c}_m + \boldsymbol{c}_R(\alpha)\\
\boldsymbol{x} \geq \boldsymbol{0}\\
\boldsymbol{y} \geq \boldsymbol{0}
\end{cases}
\end{aligned} \tag{4.6}$$

and

$$\begin{aligned}
&\min\{\boldsymbol{c}_l^{\mathrm{T}}\boldsymbol{s}\}\\
&\min\{\boldsymbol{c}_m^{\mathrm{T}}\boldsymbol{s}\}\\
&\min\{\boldsymbol{c}_r^{\mathrm{T}}\boldsymbol{s}\}\\
&\text{s.t.}\begin{cases}(\boldsymbol{B}_L(\alpha)+4\boldsymbol{B}_m+\boldsymbol{B}_R(\alpha))\boldsymbol{s}\geq(\boldsymbol{A}_L(\alpha)+4\boldsymbol{A}_m+\boldsymbol{A}_R(\alpha))\boldsymbol{z}\\(\boldsymbol{E}_L(\alpha)+4\boldsymbol{E}_m+\boldsymbol{E}_R(\alpha))\boldsymbol{z}\geq\boldsymbol{d}_L(\alpha)+4\boldsymbol{d}_m+\boldsymbol{d}_R(\alpha)\\\boldsymbol{s}\geq\boldsymbol{0}\\\boldsymbol{z}\geq\boldsymbol{0},\end{cases}
\end{aligned}\tag{4.7}$$

respectively.

Obviously, Eqs. (4.6) and (4.7) are multi-objective non-linear parameterized programming. However, if the parameter α is initially given by the players, then they are linear and are easily solved by using the simplex method of linear programming. Thus, we can provide the players a solution table with $\alpha=0,0.1,0.2,\ldots,1.0$.

Secondly, we determine the positive ideal solution and negative ideal solution of Eq. (4.6). According to Eq. (4.6), we solve the mathematical programming model as follows:

$$\begin{aligned}
&\max\{\boldsymbol{d}_l^{\mathrm{T}}\boldsymbol{x}\}\\
&\text{s.t.}\begin{cases}(\boldsymbol{E}_L(\alpha)+4\boldsymbol{E}_m+\boldsymbol{E}_R(\alpha))^{\mathrm{T}}\boldsymbol{x}\leq(\boldsymbol{A}_L(\alpha)+4\boldsymbol{A}_m+\boldsymbol{A}_R(\alpha))^{\mathrm{T}}\boldsymbol{y}\\(\boldsymbol{B}_L(\alpha)+4\boldsymbol{B}_m+\boldsymbol{B}_R(\alpha))^{\mathrm{T}}\boldsymbol{y}\leq\boldsymbol{c}_L(\alpha)+4\boldsymbol{c}_m+\boldsymbol{c}_R(\alpha)\\\boldsymbol{x}\geq\boldsymbol{0}\\\boldsymbol{y}\geq\boldsymbol{0},\end{cases}
\end{aligned}$$

denoted its optimal solution by $(\boldsymbol{y}^{1+},\boldsymbol{x}^{1+})^{\mathrm{T}}$. Thus, its optimal objective value is denoted by $D_l^+=\boldsymbol{d}_l^{\mathrm{T}}\boldsymbol{x}^{1+}$.

Analogously, according to Eq. (4.6), we solve the mathematical programming model as follows:

$$\begin{aligned}
&\max\{\boldsymbol{d}_m^{\mathrm{T}}\boldsymbol{x}\}\\
&\text{s.t.}\begin{cases}(\boldsymbol{E}_L(\alpha)+4\boldsymbol{E}_m+\boldsymbol{E}_R(\alpha))^{\mathrm{T}}\boldsymbol{x}\leq(\boldsymbol{A}_L(\alpha)+4\boldsymbol{A}_m+\boldsymbol{A}_R(\alpha))^{\mathrm{T}}\boldsymbol{y}\\(\boldsymbol{B}_L(\alpha)+4\boldsymbol{B}_m+\boldsymbol{B}_R(\alpha))^{\mathrm{T}}\boldsymbol{y}\leq\boldsymbol{c}_L(\alpha)+4\boldsymbol{c}_m+\boldsymbol{c}_R(\alpha)\\\boldsymbol{x}\geq\boldsymbol{0}\\\boldsymbol{y}\geq\boldsymbol{0},\end{cases}
\end{aligned}$$

denoted its optimal solution by $(\boldsymbol{y}^{2+},\boldsymbol{x}^{2+})^{\mathrm{T}}$. Correspondingly, its optimal objective value is denoted by $D_m^+=\boldsymbol{d}_m^{\mathrm{T}}\boldsymbol{x}^{2+}$.

We solve the mathematical programming model as follows:

$$\max\{\boldsymbol{d}_r^{\mathrm{T}}\boldsymbol{x}\}$$
$$\text{s.t.}\begin{cases}(\boldsymbol{E}_L(\alpha)+4\boldsymbol{E}_m+\boldsymbol{E}_R(\alpha))^{\mathrm{T}}\boldsymbol{x}\le(\boldsymbol{A}_L(\alpha)+4\boldsymbol{A}_m+\boldsymbol{A}_R(\alpha))^{\mathrm{T}}\boldsymbol{y}\\(\boldsymbol{B}_L(\alpha)+4\boldsymbol{B}_m+\boldsymbol{B}_R(\alpha))^{\mathrm{T}}\boldsymbol{y}\le\boldsymbol{c}_L(\alpha)+4\boldsymbol{c}_m+\boldsymbol{c}_R(\alpha)\\\boldsymbol{x}\ge\boldsymbol{0}\\\boldsymbol{y}\ge\boldsymbol{0},\end{cases}$$

denoted its optimal solution by $(\boldsymbol{y}^{3+},\boldsymbol{x}^{3+})^{\mathrm{T}}$. As a result, its optimal objective value is denoted by $D_r^+=\boldsymbol{d}_r^{\mathrm{T}}\boldsymbol{x}^{3+}$.

By computing, we have

$$D_l^-=\min\{\boldsymbol{d}_l^{\mathrm{T}}\boldsymbol{x}^{t+}\,|t=1,2,3\},$$
$$D_m^-=\min\{\boldsymbol{d}_m^{\mathrm{T}}\boldsymbol{x}^{t+}\,|t=1,2,3\}$$

and

$$D_r^-=\min\{\boldsymbol{d}_r^{\mathrm{T}}\boldsymbol{x}^{t+}\,|t=1,2,3\}.$$

Then, the positive ideal solution and negative ideal solutions of Eq. (4.6) are defined as $\boldsymbol{D}^+=(D_l^+,D_m^+,D_r^+)$ and $\boldsymbol{D}^-=(D_l^-,D_m^-,D_r^-)$, respectively.

Hereby, the relative membership functions of the three objective functions in Eq. (4.6) can be defined as follows:

$$\eta_l(\boldsymbol{d}_l^{\mathrm{T}}\boldsymbol{x})=\begin{cases}1 & \text{if}\ \ \boldsymbol{d}_l^{\mathrm{T}}\boldsymbol{x}\ge D_l^+\\\frac{\boldsymbol{d}_l^{\mathrm{T}}\boldsymbol{x}-D_l^-}{D_l^+-D_l^-} & \text{if}\ \ D_l^-\le\boldsymbol{d}_l^{\mathrm{T}}\boldsymbol{x}<D_l^+\\0 & \text{if}\ \ \boldsymbol{d}_l^{\mathrm{T}}\boldsymbol{x}<D_l^-,\end{cases}$$

$$\eta_m(\boldsymbol{d}_m^{\mathrm{T}}\boldsymbol{x})=\begin{cases}1 & \text{if}\ \ \boldsymbol{d}_m^{\mathrm{T}}\boldsymbol{x}\ge D_m^+\\\frac{\boldsymbol{d}_m^{\mathrm{T}}\boldsymbol{x}-D_m^-}{D_m^+-D_m^-} & \text{if}\ \ D_m^-\le\boldsymbol{d}_m^{\mathrm{T}}\boldsymbol{x}<D_m^+\\0 & \text{if}\ \ \boldsymbol{d}_m^{\mathrm{T}}\boldsymbol{x}<D_m^-\end{cases}$$

and

$$\eta_r(\boldsymbol{d}_r^{\mathrm{T}}\boldsymbol{x})=\begin{cases}1 & \text{if}\ \ \boldsymbol{d}_r^{\mathrm{T}}\boldsymbol{x}\ge D_r^+\\\frac{\boldsymbol{d}_r^{\mathrm{T}}\boldsymbol{x}-D_r^-}{D_r^+-D_r^-} & \text{if}\ \ D_r^-\le\boldsymbol{d}_r^{\mathrm{T}}\boldsymbol{x}<D_r^+\\0 & \text{if}\ \ \boldsymbol{d}_r^{\mathrm{T}}\boldsymbol{x}<D_r^-,\end{cases}$$

respectively.

Let

$$\eta = \min\{\lambda_l \eta_l(\boldsymbol{d}_l^{\mathrm{T}}\boldsymbol{x}), \lambda_m \eta_m(\boldsymbol{d}_m^{\mathrm{T}}\boldsymbol{x}), \lambda_r \eta_r(\boldsymbol{d}_r^{\mathrm{T}}\boldsymbol{x})\},$$

where $\lambda_l \geq 0$, $\lambda_m \geq 0$, and $\lambda_r \geq 0$ are weights, and $\lambda_l + \lambda_m + \lambda_r = 1$. Using Zimmermann's fuzzy programming method [7], hence, Eq. (4.6) can be aggregated into the mathematical programming model as follows:

$$\max\{\eta\}$$
$$\text{s.t.}\begin{cases} (\boldsymbol{E}_L(\alpha) + 4\boldsymbol{E}_m + \boldsymbol{E}_R(\alpha))^{\mathrm{T}}\boldsymbol{x} \leq (\boldsymbol{A}_L(\alpha) + 4\boldsymbol{A}_m + \boldsymbol{A}_R(\alpha))^{\mathrm{T}}\boldsymbol{y} \\ (\boldsymbol{B}_L(\alpha) + 4\boldsymbol{B}_m + \boldsymbol{B}_R(\alpha))^{\mathrm{T}}\boldsymbol{y} \leq \boldsymbol{c}_L(\alpha) + 4\boldsymbol{c}_m + \boldsymbol{c}_R(\alpha) \\ \lambda_l(\boldsymbol{d}_l^{\mathrm{T}}\boldsymbol{x} - D_l^-) \geq \eta(D_l^+ - D_l^-) \\ \lambda_m(\boldsymbol{d}_m^{\mathrm{T}}\boldsymbol{x} - D_m^-) \geq \eta(D_m^+ - D_m^-) \\ \lambda_r(\boldsymbol{d}_r^{\mathrm{T}}\boldsymbol{x} - D_r^-) \geq \eta(D_r^+ - D_r^-) \\ \boldsymbol{x} \geq \boldsymbol{0} \\ \boldsymbol{y} \geq \boldsymbol{0} \\ 0 \leq \eta \leq 1, \end{cases} \tag{4.8}$$

where η is a decision variable, $\boldsymbol{x}$ and $\boldsymbol{y}$ the vectors of decision variables, $\alpha \in [0, 1]$ is a parameter. As stated earlier, if the parameter $\alpha \in [0, 1]$ is known a priori, then Eq. (4.8) is linear programming and hereby can be easily solved by using the simplex method of linear programming.

Solving Eq. (4.8), we can obtain its optimal solution $(\boldsymbol{y}^*, \boldsymbol{x}^*, \eta^*)^{\mathrm{T}}$. Thus, we obtain the optimal or maximin (mixed) strategy $\boldsymbol{y}^*$ and gain-floor $\tilde{\upsilon}^* = \tilde{\boldsymbol{d}}^{\mathrm{T}}\boldsymbol{x}^*$ for the player I.

In the same way as previously described, according to Eq. (4.7), we solve the mathematical programming model as follows:

$$\min\{\boldsymbol{c}_l^{\mathrm{T}}\boldsymbol{s}\}$$
$$\text{s.t.}\begin{cases} (\boldsymbol{B}_L(\alpha) + 4\boldsymbol{B}_m + \boldsymbol{B}_R(\alpha))\boldsymbol{s} \geq (\boldsymbol{A}_L(\alpha) + 4\boldsymbol{A}_m + \boldsymbol{A}_R(\alpha))\boldsymbol{z} \\ (\boldsymbol{E}_L(\alpha) + 4\boldsymbol{E}_m + \boldsymbol{E}_R(\alpha))\boldsymbol{z} \geq \boldsymbol{d}_L(\alpha) + 4\boldsymbol{d}_m + \boldsymbol{d}_R(\alpha) \\ \boldsymbol{s} \geq \boldsymbol{0} \\ \boldsymbol{z} \geq \boldsymbol{0}, \end{cases}$$

denoted its optimal solution by $(\boldsymbol{z}^{1+}, \boldsymbol{s}^{1+})^{\mathrm{T}}$. As a result, its optimal objective value is denoted by $G_l^+ = \boldsymbol{c}_l^{\mathrm{T}}\boldsymbol{s}^{1+}$.

Analogously, according to Eq. (4.7), we solve the mathematical programming model as follows:

$$\min\{\boldsymbol{c}_l^{\mathrm{T}}\boldsymbol{s}\}$$
$$\text{s.t.}\begin{cases} (\boldsymbol{B}_L(\alpha) + 4\boldsymbol{B}_m + \boldsymbol{B}_R(\alpha))\boldsymbol{s} \geq (\boldsymbol{A}_L(\alpha) + 4\boldsymbol{A}_m + \boldsymbol{A}_R(\alpha))\boldsymbol{z} \\ (\boldsymbol{E}_L(\alpha) + 4\boldsymbol{E}_m + \boldsymbol{E}_R(\alpha))\boldsymbol{z} \geq \boldsymbol{d}_L(\alpha) + 4\boldsymbol{d}_m + \boldsymbol{d}_R(\alpha) \\ \boldsymbol{s} \geq \boldsymbol{0} \\ \boldsymbol{z} \geq \boldsymbol{0}, \end{cases}$$

denoted its optimal solution by $(z^{2+}, s^{2+})^{\mathrm{T}}$. Consequently, its optimal objective value is denoted by $G_m^+ = \boldsymbol{c}_m^{\mathrm{T}} \boldsymbol{s}^{2+}$.

According to Eq. (4.7), we solve the mathematical programming model as follows:

$$\min\{\boldsymbol{c}_r^{\mathrm{T}} \boldsymbol{s}\}$$
$$\text{s.t.} \begin{cases} (\boldsymbol{B}_L(\alpha) + 4\boldsymbol{B}_m + \boldsymbol{B}_R(\alpha))\boldsymbol{s} \geq (\boldsymbol{A}_L(\alpha) + 4\boldsymbol{A}_m + \boldsymbol{A}_R(\alpha))\boldsymbol{z} \\ (\boldsymbol{E}_L(\alpha) + 4\boldsymbol{E}_m + \boldsymbol{E}_R(\alpha))\boldsymbol{z} \geq \boldsymbol{d}_L(\alpha) + 4\boldsymbol{d}_m + \boldsymbol{d}_R(\alpha) \\ \boldsymbol{s} \geq \boldsymbol{0} \\ \boldsymbol{z} \geq \boldsymbol{0}, \end{cases}$$

denoted its optimal solution by $(z^{3+}, s^{3+})^{\mathrm{T}}$. Consequently, its optimal objective value is denoted by $G_r^+ = \boldsymbol{c}_r^{\mathrm{T}} \boldsymbol{s}^{3+}$.

We compute

$$G_l^- = \max\{\boldsymbol{c}_l^{\mathrm{T}} \boldsymbol{s}^{t+} \mid t = 1, 2, 3\},$$
$$G_m^- = \max\{\boldsymbol{c}_m^{\mathrm{T}} \boldsymbol{s}^{t+} \mid t = 1, 2, 3\}$$

and

$$G_r^- = \max\{\boldsymbol{c}_r^{\mathrm{T}} \boldsymbol{s}^{t+} \mid t = 1, 2, 3\}.$$

Then, the positive ideal solution and negative ideal solutions of Eq. (4.7) are defined as $\boldsymbol{G}^+ = (G_l^+, G_m^+, G_r^+)$ and $\boldsymbol{G}^- = (G_l^-, G_m^-, G_r^-)$, respectively.

The relative membership functions of the three objective functions in Eq. (4.7) can be defined as follows:

$$\rho_l(\boldsymbol{c}_l^{\mathrm{T}} \boldsymbol{s}) = \begin{cases} 1 & \text{if } \boldsymbol{c}_l^{\mathrm{T}} \boldsymbol{s} \leq G_l^+ \\ \frac{\boldsymbol{c}_l^{\mathrm{T}} \boldsymbol{s} - G_l^+}{G_l^- - G_l^+} & \text{if } G_l^+ < \boldsymbol{c}_l^{\mathrm{T}} \boldsymbol{s} \leq G_l^- \\ 0 & \text{if } \boldsymbol{c}_l^{\mathrm{T}} \boldsymbol{s} > G_l^-, \end{cases}$$

$$\rho_m(\boldsymbol{c}_m^{\mathrm{T}} \boldsymbol{s}) = \begin{cases} 1 & \text{if } \boldsymbol{c}_m^{\mathrm{T}} \boldsymbol{s} \leq G_m^+ \\ \frac{\boldsymbol{c}_m^{\mathrm{T}} \boldsymbol{s} - G_m^+}{G_m^- - G_m^+} & \text{if } G_m^+ < \boldsymbol{c}_m^{\mathrm{T}} \boldsymbol{s} \leq G_m^- \\ 0 & \text{if } \boldsymbol{c}_m^{\mathrm{T}} \boldsymbol{s} > G_m^- \end{cases}$$

and

$$\rho_r(\boldsymbol{c}_r^{\mathrm{T}} \boldsymbol{s}) = \begin{cases} 1 & \text{if } \boldsymbol{c}_r^{\mathrm{T}} \boldsymbol{s} \leq G_r^+ \\ \frac{\boldsymbol{c}_r^{\mathrm{T}} \boldsymbol{s} - G_r^+}{G_r^- - G_r^+} & \text{if } G_r^+ < \boldsymbol{c}_r^{\mathrm{T}} \boldsymbol{s} \leq G_r^- \\ 0 & \text{if } \boldsymbol{c}_r^{\mathrm{T}} \boldsymbol{s} > G_r^-, \end{cases}$$

respectively.

Let

$$\rho = \min\{\theta_l \rho_l(\boldsymbol{c}_l^{\mathrm{T}}\boldsymbol{s}), \theta_m \rho_m(\boldsymbol{c}_m^{\mathrm{T}}\boldsymbol{s}), \theta_r \rho_r(\boldsymbol{c}_r^{\mathrm{T}}\boldsymbol{s})\},$$

where $\theta_l \geq 0$, $\theta_m \geq 0$, and $\theta_r \geq 0$ are weights, and $\theta_l + \theta_m + \theta_r = 1$.

Using Zimmermann's fuzzy programming method [7], hence, Eq. (4.7) can be aggregated into the mathematical programming model as follows:

$$\min\{\rho\}$$
$$\text{s.t.} \begin{cases} (\boldsymbol{B}_L(\alpha) + 4\boldsymbol{B}_m + \boldsymbol{B}_R(\alpha))\boldsymbol{s} \geq (\boldsymbol{A}_L(\alpha) + 4\boldsymbol{A}_m + \boldsymbol{A}_R(\alpha))\boldsymbol{z} \\ (\boldsymbol{E}_L(\alpha) + 4\boldsymbol{E}_m + \boldsymbol{E}_R(\alpha))\boldsymbol{z} \geq \boldsymbol{d}_L(\alpha) + 4\boldsymbol{d}_m + \boldsymbol{d}_R(\alpha) \\ \theta_l(\boldsymbol{c}_l^{\mathrm{T}}\boldsymbol{s} - G_l^+) \geq \rho(G_l^- - G_l^+) \\ \theta_m(\boldsymbol{c}_m^{\mathrm{T}}\boldsymbol{s} - G_m^+) \geq \rho(G_m^- - G_m^+) \\ \theta_r(\boldsymbol{c}_r^{\mathrm{T}}\boldsymbol{s} - G_r^+) \geq \rho(G_r^- - G_r^+) \\ \boldsymbol{s} \geq \boldsymbol{0} \\ \boldsymbol{z} \geq \boldsymbol{0} \\ 0 \leq \rho \leq 1, \end{cases} \tag{4.9}$$

where ρ is a decision variable, $\boldsymbol{s}$ and $\boldsymbol{z}$ the vectors of decision variables.

Solving Eq. (4.9), we can obtain its optimal solution $(\boldsymbol{z}^*, \boldsymbol{s}^*, \rho^*)^{\mathrm{T}}$. Hence, we obtain the optimal or minimax (mixed) strategy $\boldsymbol{z}^*$ and loss-ceiling $\tilde{\omega}^* = \tilde{\boldsymbol{c}}^{\mathrm{T}}\boldsymbol{s}^*$ for the player II.

Example 4.1 Let us consider a simple constrained matrix game $\tilde{\boldsymbol{A}}_0$ with payoffs of triangular fuzzy numbers. Assume that the fuzzy payoff matrix of the player I is given as follows:

$$\tilde{\boldsymbol{A}}_0 = \begin{matrix} \\ \delta_1 \\ \delta_2 \end{matrix} \begin{matrix} \beta_1 & \beta_2 \\ \left(\begin{matrix} (18, 20, 23) \\ (-33, -32, -27) \end{matrix} \right. & \left. \begin{matrix} (-21, -18, -16) \\ (38, 40, 43) \end{matrix} \right) \end{matrix}.$$

The coefficient matrices and vectors of the constraint sets of strategies for the player I and II are expressed as follows:

$$\tilde{\boldsymbol{B}}_0 = \begin{pmatrix} (70, 80, 88) & 1 & -1 \\ (44, 50, 54) & 1 & -1 \end{pmatrix},$$
$$\tilde{\boldsymbol{E}}_0^{\mathrm{T}} = \begin{pmatrix} (-48, -40, -35) & 1 & -1) \\ (-79, -70, -65) & 1 & -1) \end{pmatrix},$$
$$\tilde{c}_0 = ((61, 67, 74), (0, 1, 0), (0, -1, 0))^{\mathrm{T}}$$

and

$$\tilde{\boldsymbol{d}}_0 = ((-60, -52, -50), (0, 1, 0), (0, -1, 0))^{\mathrm{T}},$$

respectively.

Taking $\alpha = 0.5$, $\lambda_l = \lambda_m = \lambda_r = 1/3$, and $\theta_l = \theta_m = \theta_r = 1/3$, and according to Eqs. (4.8) and (4.9), by using the simplex method of linear programming, we obtain the optimal (or maximin) mixed strategy $\boldsymbol{y}_0^*$ and gain-floor $\tilde{\upsilon}_0^* = \tilde{\boldsymbol{d}}_0^{\mathrm{T}} \boldsymbol{x}_0^*$ for the player I and the optimal (or minimax) mixed strategy $\boldsymbol{z}_0^*$ and loss-ceiling $\tilde{\omega}_0^* = \tilde{\boldsymbol{c}}_0^{\mathrm{T}} \boldsymbol{s}_0^*$ for the player II, where

$$\boldsymbol{y}_0^* = (0.575, 0.425)^{\mathrm{T}},$$
$$\tilde{\upsilon}_0^* = (1.02, 3.292, 3.86),$$
$$\boldsymbol{z}_0^* = (0.407, 0.593)^{\mathrm{T}}$$

and

$$\tilde{\omega}_0^* = (2.448, 5.136, 8.272),$$

respectively.

Furthermore, according to Eq. (4.3), we obtain the fuzzy equilibrium value $\tilde{V}_0^*$ of the constrained matrix game $\tilde{\boldsymbol{A}}_0$ with payoffs of triangular fuzzy numbers, where

$$\mu_{\tilde{V}_0^*}(x) = \begin{cases} \frac{x-2.448}{5.136-2.448} & \text{if} \quad 2.448 \le x < 3.613 \\ 0.433 & \text{if} \quad x = 3.613 \\ \frac{3.86-x}{3.86-3.292} & \text{if} \quad 3.613 < x \le 3.86 \\ 0 & \text{else}, \end{cases}$$

i.e.,

$$\mu_{\tilde{V}_0^*}(x) = \begin{cases} 0.372x - 0.911 & \text{if} \quad 2.448 \le x < 3.613 \\ 0.433 & \text{if} \quad x = 3.613 \\ 6.795 - 1.761x & \text{if} \quad 3.613 < x \le 3.86 \\ 0 & \text{else}, \end{cases}$$

depicted as in Fig. 4.1. Therefore, there exists a fuzzy equilibrium value 3.613 with possibility of 0.433. In other words, the fuzzy value of the constrained matrix game $\tilde{\boldsymbol{A}}_0$ with payoffs of triangular fuzzy numbers is "around 3.613". Or the player I's minimum reward is 2.448 while his/her maximum reward is 3.86. He/she can win any intermediate value x between 2.448 and 3.86 with the possibility $\mu_{\tilde{V}_0^*}(x)$.

If other values of λ_l, λ_m, λ_r, θ_l, θ_m, and θ_r are chosen, then we can similarly solve corresponding Eqs. (4.8) and (4.9) and hereby obtain the players' optimal mixed strategies, the player I's gain-floor, and the player II's loss-ceiling (omitted).

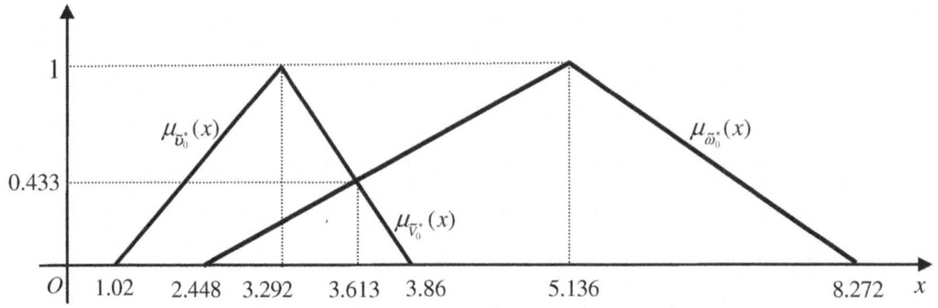

Fig. 4.1 The fuzzy equilibrium value $\tilde{V}_0^*$

4.3 Alfa-Cut-Based Primal-Dual Linear Programming Models of Constrained Matrix Games with Payoffs of Triangular Fuzzy Numbers

4.3.1 Concepts of Alfa-Constrained Matrix Games with Payoffs of Triangular Fuzzy Numbers

Let us continue to consider the constrained matrix game $\tilde{\boldsymbol{A}}$ with payoffs of triangular fuzzy numbers, where the payoff matrix of the player I is given as $\tilde{\boldsymbol{A}} = (\tilde{a}_{ij})_{m\times n}$, whose elements $\tilde{a}_{ij}$ $(i = 1, 2, \ldots, m;\ j = 1, 2, \ldots, n)$ are triangular fuzzy numbers stated as in Sect. 2.2, the sets of pure strategies and the constraint sets of mixed strategies for the player I and the player II are $S_1 = \{\delta_1, \delta_2, \ldots, \delta_m\}$, $S_2 = \{\beta_1, \beta_2, \ldots, \beta_n\}$, $Y = \{\boldsymbol{y}|\boldsymbol{B}^{\mathrm{T}}\boldsymbol{y} \le \boldsymbol{c}, \boldsymbol{y} \ge \boldsymbol{0}\}$, and $Z = \{\boldsymbol{z}|\boldsymbol{E}\boldsymbol{z} \ge \boldsymbol{d}, \boldsymbol{z} \ge \boldsymbol{0}\}$ stated as in Sect. 3.2, respectively.

For any $\alpha \in [0, 1]$, denote the payoff matrix of the player I by $\tilde{\boldsymbol{A}}(\alpha) = (\tilde{a}_{ij}(\alpha))_{m\times n}$, whose elements $\tilde{a}_{ij}(\alpha)$ $(i = 1, 2, \ldots, m;\ j = 1, 2, \ldots, n)$ are α-cuts of the payoffs $\tilde{a}_{ij}$ which are triangular fuzzy numbers. Then, $\tilde{\boldsymbol{A}}(\alpha)$ is called a α-constrained matrix game corresponding to the constrained matrix game $\tilde{\boldsymbol{A}}$ with payoffs of triangular fuzzy numbers in the α-confidence level, which often is called the α-constrained matrix game $\tilde{\boldsymbol{A}}(\alpha)$ for short. It is noted that the sets S_1 and S_2 of pure strategies and the constraint sets Y and Z of mixed strategies for the player I and the player II in the α-constrained matrix game $\tilde{\boldsymbol{A}}(\alpha)$ are the same as those in the constrained matrix game $\tilde{\boldsymbol{A}}$ with payoffs of triangular fuzzy numbers.

Definition 4.1 For any $\alpha \in [0, 1]$, if the player I's gain-floor $\tilde{v}(\alpha)$ and the player II's loss-ceiling $\tilde{\rho}(\alpha)$ have a common value $\tilde{V}(\alpha)$, then $\tilde{V}(\alpha)$ is called a value of the α-constrained matrix game $\tilde{\boldsymbol{A}}(\alpha)$, or the α-constrained matrix game $\tilde{\boldsymbol{A}}(\alpha)$ has a value $\tilde{V}(\alpha)$, where $\tilde{V}(\alpha) = \tilde{v}(\alpha) = \tilde{\rho}(\alpha)$.

Definition 4.1 is completely the same as that of classical matrix games, which is rational since it reflects that one player wins the other player loses in matrix games.

Definition 4.2 For any $\alpha \in [0,1]$, if every α-constrained matrix game $\tilde{A}(\alpha)$ has a value $\tilde{V}(\alpha)$, then the constrained matrix game $\tilde{A}$ with payoffs of triangular fuzzy numbers has a fuzzy value $\tilde{V}$, where $\tilde{V} = \bigcup_{\alpha \in [0,1]} \{\alpha \otimes \tilde{V}(\alpha)\}$.

4.3.2 Linear Programming Models of Constrained Matrix Games with Payoffs of Triangular Fuzzy Numbers

For the constrained matrix game $\tilde{A}$ with payoffs of triangular fuzzy numbers stated as in the above Sect. 4.3.1, as stated earlier, there is no method which can always ensure that the player I's gain-floor and the player II's loss-ceiling have a common value and hereby the constrained matrix game $\tilde{A}$ with payoffs of triangular fuzzy numbers has a fuzzy value. In this subsection, inspired by our work [10], according to Definitions 4.1 and 4.2, we develop a linear programming method for solving any α-constrained matrix game.

Stated as earlier, for any $\alpha \in [0,1]$, let us consider a α-constrained matrix game $\tilde{A}(\alpha)$, where the payoff matrix of the player I is given as $\tilde{A}(\alpha) = (\tilde{a}_{ij}\alpha))_{m \times n}$, whose elements $\tilde{a}_{ij}(\alpha)$ are the α-cuts of the triangular fuzzy numbers $\tilde{a}_{ij} = (a_{ij}^l, a_{ij}^m, a_{ij}^r)$ $(i = 1, 2, \ldots, m; j = 1, 2, \ldots, n)$, i.e.,

$$\tilde{a}_{ij}(\alpha) = [a_{ij}^L(\alpha), a_{ij}^R(\alpha)] = [\alpha a_{ij}^m + (1-\alpha) a_{ij}^l, \alpha a_{ij}^m + (1-\alpha) a_{ij}^r] \tag{4.10}$$

according to Eq. (2.4) or Eq. (2.26). Essentially, the α-constrained matrix game $\tilde{A}(\alpha)$ is the interval-valued constrained matrix game, which has been discussed in Sect. 3.3.

For any given values $a_{ij}(\alpha)$ in the interval-valued payoffs $\tilde{a}_{ij}(\alpha)$ $(i = 1, 2, \ldots, m;$ $j = 1, 2, \ldots, n)$, a payoff matrix is denoted by $A(\alpha) = (a_{ij}(\alpha))_{m \times n}$. It is easily seen from Eq. (3.3) that the value $v(\alpha)$ of the constrained matrix game $A(\alpha)$ for the player I is closely related to all $a_{ij}(\alpha)$, i.e., entries in the payoff matrix $A(\alpha)$. In other words, $v(\alpha)$ is a function of the values $a_{ij}(\alpha)$ in the interval-valued payoffs $\tilde{a}_{ij}(\alpha)$, denoted by $v(\alpha) = \upsilon(a_{ij}(\alpha))$ or $v(\alpha) = \upsilon(A(\alpha))$. Similarly, the optimal mixed strategy $y^*(\alpha) \in Y$ of the player I is also a function of the values $a_{ij}(\alpha)$ $(i = 1, 2, \ldots, m;$ $j = 1, 2, \ldots, n)$, denoted by $y^*(\alpha) = y^*(\alpha_{ij}(\alpha))$ or $y^*(\alpha) = y^*(A(\alpha))$.

In a similar way to the above analysis, the value $\mu(\alpha)$ and the optimal mixed strategy $z^*(\alpha) \in Z$ for the player II in the constrained matrix game $A(\alpha)$ are functions of the values $a_{ij}(\alpha)$ $(i = 1, 2, \ldots, m; j = 1, 2, \ldots, n)$ in the interval-valued payoffs $\tilde{a}_{ij}(\alpha)$, denoted by $\mu(\alpha) = \omega(a_{ij}(\alpha))$ (or $\mu(\alpha) = \omega(A(\alpha))$) and $z^*(\alpha) = z^*(a_{ij}(\alpha))$ (or $z^*(\alpha) = z^*(A(\alpha))$), respectively.

In a similar way to the discussion in Sect. 3.3, it is proven that the gain-floor $v(\alpha) = \upsilon(a_{ij}(\alpha))$ for the player I is a monotonic and non-decreasing function of the values $a_{ij}(\alpha)$ $(i = 1, 2, \ldots, m; j = 1, 2, \ldots, n)$ in the interval-valued payoffs $\tilde{a}_{ij}(\alpha)$. In fact, for any values $a_{ij}(\alpha)$ and $a'_{ij}(\alpha)$ in the interval-valued payoffs $\tilde{a}_{ij}(\alpha)$ $(i = 1, 2, \ldots, m; j = 1, 2, \ldots, n)$, if $a_{ij}(\alpha) \leq a'_{ij}(\alpha)$, then we have

$$\sum_{i=1}^{m}\sum_{j=1}^{n} y_i a_{ij}(\alpha) z_j \leq \sum_{i=1}^{m}\sum_{j=1}^{n} y_i a'_{ij}(\alpha) z_j$$

due to $y_i \geq 0$ $(i = 1, 2, \ldots, m)$ and $z_j \geq 0$ $(j = 1, 2, \ldots, n)$, where $\mathbf{y} \in Y$ and $\mathbf{z} \in Z$. Hence, we have

$$\min_{z \in Z}\{\sum_{i=1}^{m}\sum_{j=1}^{n} y_i a_{ij}(\alpha) z_j\} \leq \min_{z \in Z}\{\sum_{i=1}^{m}\sum_{j=1}^{n} y_i a'_{ij}(\alpha) z_j\},$$

which directly implies that

$$\max_{y \in Z}\min_{z \in Z}\{\sum_{i=1}^{m}\sum_{j=1}^{n} y_i a_{ij}(\alpha) z_j\} \leq \max_{y \in Z}\min_{z \in Z}\{\sum_{i=1}^{m}\sum_{j=1}^{n} y_i a'_{ij}(\alpha) z_j\},$$

i.e., $\upsilon(a_{ij}(\alpha)) \leq \upsilon(a'_{ij}(\alpha))$ or $\upsilon(\mathbf{A}(\alpha)) \leq \upsilon(\mathbf{A}'(\alpha))$, where $\mathbf{A}'(\alpha) = (a'_{ij}(\alpha))_{m \times n}$ is the payoff matrix of the player I in the constrained matrix game $\mathbf{A}'(\alpha)$.

According to Theorems 3.1 and 3.2 or the minimax theorem of constrained matrix games [4], if the constrained matrix game $\mathbf{A}(\alpha) = (a_{ij}(\alpha))_{m \times n}$ has a value, then it is denoted by $V(\alpha) = V(a_{ij}(\alpha))$ or $V(\alpha) = V(\mathbf{A}(\alpha))$. Obviously, $V(\alpha) = v(\alpha) = \mu(\alpha)$. From the above discussion, $V(\alpha) = V(a_{ij}(\alpha))$ is also a non-decreasing function of the values $a_{ij}(\alpha)$ $(i = 1, 2, \ldots, m; j = 1, 2, \ldots, n)$ in the interval-valued payoffs $\tilde{a}_{ij}(\alpha)$.

For the α-constrained matrix game $\tilde{\mathbf{A}}(\alpha)$, the expected payoffs of the players are a linear combination of interval-valued payoffs. Thus, from a viewpoint of logic and interval operations, the value of the α-constrained matrix game $\tilde{\mathbf{A}}(\alpha)$ should be a closed interval as well [10, 11]. Stated as earlier, the value $v(\alpha) = \upsilon(a_{ij}(\alpha))$ of the constrained matrix game $\mathbf{A}(\alpha) = (a_{ij}(\alpha))_{m \times n}$ for the player I is a non-decreasing function of the values $a_{ij}(\alpha)$ $(i = 1, 2, \ldots, m; j = 1, 2, \ldots, n)$ in the interval-valued payoffs $\tilde{a}_{ij}(\alpha)$. Hence, the upper bound $v^R(\alpha)$ of the interval-type value $\tilde{v}(\alpha)$ of the α-constrained matrix game $\tilde{\mathbf{A}}(\alpha)$ for the player I can be obtained as follows:

$$v^R(\alpha) = \max_{y \in Y}\min_{z \in Z}\{\mathbf{y}^{\mathrm{T}}\mathbf{A}^R(\alpha)\mathbf{z}\} = \max_{y \in Y}\min_{z \in Z}\{\sum_{i=1}^{m}\sum_{j=1}^{n} y_i a^R_{ij}(\alpha) z_j\}, \tag{4.11}$$

where $\boldsymbol{A}^R(\alpha)=(a_{ij}^R(\alpha))_{m\times n}$. According to Eq. (3.5), (4.11) is equivalent to the linear programming model as follows:

$$\max\{\boldsymbol{d}^{\mathrm{T}}\boldsymbol{x}^R(\alpha)\}$$
$$\text{s.t.}\begin{cases}\boldsymbol{E}^{\mathrm{T}}\boldsymbol{x}^R(\alpha)-(\boldsymbol{A}^R(\alpha))^{\mathrm{T}}\boldsymbol{y}^R(\alpha)\leq 0\\ \boldsymbol{B}^{\mathrm{T}}\boldsymbol{y}^R(\alpha)\leq \boldsymbol{c}\\ \boldsymbol{x}^R(\alpha)\geq 0\\ \boldsymbol{y}^R(\alpha)\geq 0,\end{cases} \tag{4.12}$$

where $\boldsymbol{x}^R(\alpha)$ and $\boldsymbol{y}^R(\alpha)$ are vectors of decision variables.

If Eq. (4.12) is feasible linear programming, then by using the simplex method of linear programming [4], we can obtain its optimal solution, denoted by $(\boldsymbol{x}^{R*}(\alpha),\boldsymbol{y}^{R*}(\alpha))$. Thus, according to Theorem 3.3, we obtain the upper bound $v^R(\alpha)=\boldsymbol{d}^{\mathrm{T}}\boldsymbol{x}^{R*}(\alpha)$ of the player I's gain-floor $\tilde{v}(\alpha)$ and corresponding optimal mixed strategy $\boldsymbol{y}^{R*}(\alpha)$ for the α-constrained matrix game $\tilde{\boldsymbol{A}}(\alpha)$.

Analogously, the lower bound $v^L(\alpha)$ of the player I's gain-floor $\tilde{v}(\alpha)$ and the optimal mixed strategy $\boldsymbol{y}^{L*}(\alpha)\in Y$ for the α-constrained matrix game $\tilde{\boldsymbol{A}}(\alpha)$ are $v^L(\alpha)=\upsilon(a_{ij}^L(\alpha))$ and $\boldsymbol{y}^{L*}(\alpha)=\boldsymbol{y}^*(a_{ij}^L(\alpha))$, respectively. According to Eq. (3.5), $(v^L(\alpha),\boldsymbol{y}^{L*}(\alpha))$ can be obtained by solving the linear programming model as follows:

$$\max\{\boldsymbol{d}^{\mathrm{T}}\boldsymbol{x}^L(\alpha)\}$$
$$\text{s.t.}\begin{cases}\boldsymbol{E}^{\mathrm{T}}\boldsymbol{x}^L(\alpha)-(\boldsymbol{A}^L(\alpha))^{\mathrm{T}}\boldsymbol{y}^L(\alpha)\leq 0\\ \boldsymbol{B}^{\mathrm{T}}\boldsymbol{y}^L(\alpha)\leq \boldsymbol{c}\\ \boldsymbol{x}^L(\alpha)\geq 0\\ \boldsymbol{y}^L(\alpha)\geq 0,\end{cases} \tag{4.13}$$

where $\boldsymbol{x}^L(\alpha)$ and $\boldsymbol{y}^L(\alpha)$ are vectors of decision variables, and $\boldsymbol{A}^L(\alpha)=(a_{ij}^L(\alpha))_{m\times n}$.

If Eq. (4.13) is feasible linear programming, then we can obtain its optimal solution, denoted by $(\boldsymbol{y}^{L*}(\alpha),\boldsymbol{x}^{L*}(\alpha))^{\mathrm{T}}$. Thus, according to Theorem 3.2, we obtain the lower bound $\mu^L(\alpha)=\boldsymbol{c}^{\mathrm{T}}\boldsymbol{s}^{L*}(\alpha)$ of the player I's gain-floor $\tilde{v}(\alpha)$ and corresponding optimal mixed strategy $\boldsymbol{y}^{L*}(\alpha)$ for the α-constrained matrix game $\tilde{\boldsymbol{A}}(\alpha)$.

Thus, the lower bound $v^L(\alpha)$ and the upper bound $v^R(\alpha)$ of the interval-type value $\tilde{v}(\alpha)$ of the α-constrained matrix game $\tilde{\boldsymbol{A}}(\alpha)$ for the player I can be obtained. Therefore, the value of the α-constrained matrix game $\tilde{\boldsymbol{A}}(\alpha)$ is a closed interval $[v^L(\alpha),v^R(\alpha)]$. Namely, $\tilde{v}(\alpha)=[v^L(\alpha),v^R(\alpha)]$. It is obvious that $\tilde{v}(\alpha)$ is a α-cut of the player I's gain-floor $\tilde{v}$ in the constrained matrix game $\tilde{\boldsymbol{A}}$ with payoffs of triangular fuzzy numbers.

In the same analysis to that of the player I, the upper bound $\mu^R(\alpha)$ of the interval-type value $\tilde{\mu}(\alpha)$ of the α-constrained matrix game $\tilde{\boldsymbol{A}}(\alpha)$ and corresponding optimal mixed strategy $z^{R*}(\alpha) \in Z$ for the player II are $\mu^R(\alpha) = \omega(a_{ij}^R(\alpha))$ and $z^{R*}(\alpha) = z^*(a_{ij}^R(\alpha))$, respectively. According to Eq. (3.6), $(\mu^R(\alpha), z^{R*}(\alpha))$ can be obtained by solving the linear programming model as follows:

$$\min\{\boldsymbol{c}^{\mathrm{T}}\boldsymbol{s}^R(\alpha)\}$$
$$\text{s.t.}\begin{cases} \boldsymbol{B}\boldsymbol{s}^R(\alpha) - \boldsymbol{A}^R(\alpha)\boldsymbol{z}^R(\alpha) \geq 0 \\ \boldsymbol{E}\boldsymbol{z}^R(\alpha) \geq \boldsymbol{d} \\ \boldsymbol{s}^R(\alpha) \geq 0 \\ \boldsymbol{z}^R(\alpha) \geq 0, \end{cases} \tag{4.14}$$

where $\boldsymbol{s}^R(\alpha)$ and $\boldsymbol{z}^R(\alpha)$ are vectors of decision variables.

If Eq. (4.14) is feasible linear programming, then by using the simplex method of linear programming, we can obtain its optimal solution, denoted by $(\boldsymbol{z}^{R*}(\alpha), \boldsymbol{s}^{R*}(\alpha))$. Thus, according to Theorem 3.2, we obtain the upper bound $\mu^R(\alpha) = \boldsymbol{c}^{\mathrm{T}}\boldsymbol{s}^{R*}(\alpha)$ of the player II's loss-ceiling $\tilde{\mu}(\alpha)$ and corresponding optimal mixed strategy $\boldsymbol{z}^{R*}(\alpha)$.

In the same way, the lower bound $\mu^L(\alpha)$ of the player II's loss-ceiling $\tilde{\mu}(\alpha)$ and corresponding optimal mixed strategy $z^{L*}(\alpha) \in Z$ are $\mu^L(\alpha) = \omega(a_{ij}^L(\alpha))$ and $z^{L*}(\alpha) = z^*(a_{ij}^L(\alpha))$, respectively. According to Eq. (3.6), $(\omega^L(\alpha), z^{L*}(\alpha))$ can be obtained by solving the linear programming model as follows:

$$\min\{\boldsymbol{c}^{\mathrm{T}}\boldsymbol{s}^L(\alpha)\}$$
$$\text{s.t.}\begin{cases} \boldsymbol{B}\boldsymbol{s}^L(\alpha) - \boldsymbol{A}^L(\alpha)\boldsymbol{z}^L(\alpha) \geq 0 \\ \boldsymbol{E}\boldsymbol{z}^L(\alpha) \geq \boldsymbol{d} \\ \boldsymbol{s}^L(\alpha) \geq 0 \\ \boldsymbol{z}^L(\alpha) \geq 0, \end{cases} \tag{4.15}$$

where $\boldsymbol{s}^L(\alpha)$ and $\boldsymbol{z}^L(\alpha)$ are vectors of decision variables.

If Eq. (4.15) is feasible linear programming, then we can obtain its optimal solution, denoted by $(\boldsymbol{z}^{L*}(\alpha), \boldsymbol{s}^{L*}(\alpha))^{\mathrm{T}}$. Thus, we obtain the lower bound $\mu^L(\alpha) = \boldsymbol{c}^{\mathrm{T}}\boldsymbol{s}^{L*}(\alpha)$ of the player II's loss-ceiling $\tilde{\mu}(\alpha)$ and optimal mixed strategy $\boldsymbol{z}^{L*}(\alpha)$ in the α-constrained matrix game $\tilde{\boldsymbol{A}}(\alpha)$.

Thus, the lower bound $\mu^L(\alpha)$ and the upper bound $\mu^R(\alpha)$ of the interval-type value $\tilde{\mu}(\alpha)$ of the α-constrained matrix game $\tilde{\boldsymbol{A}}(\alpha)$ for the player II can be obtained. Therefore, the player II's loss-ceiling of the α-constrained matrix game $\tilde{\boldsymbol{A}}(\alpha)$ is a closed interval $[\mu^L(\alpha), \mu^R(\alpha)]$. Namely, $\tilde{\mu}(\alpha) = [\mu^L(\alpha), \mu^R(\alpha)]$. It is obvious that

$\tilde{\mu}(\alpha)$ is the α-cut of the player II's loss-ceiling $\tilde{\mu}$ in the constrained matrix game $\tilde{\boldsymbol{A}}$ with payoffs of triangular fuzzy numbers.

It is easy to see that Eqs. (4.12) and (4.14) are a pair of primal-dual linear programming models. Thus, the maximum of $\upsilon^R(\alpha)$ is equal to the minimum of $\omega^R(\alpha)$ according to the duality theorem of linear programming [4], i.e.,

$$v^R(\alpha) = \mu^R(\alpha). \tag{4.16}$$

Likewise, Eqs. (4.13) and (4.15) are a pair of primal-dual linear programming models. Hence, we have

$$v^L(\alpha) = \mu^L(\alpha). \tag{4.17}$$

Therefore, the player I's gain-floor $\tilde{v}(\alpha) = [v^L(\alpha), v^R(\alpha)]$ is equal to the player II's loss-ceiling $\tilde{\mu}(\alpha) = [\mu^L(\alpha), \mu^R(\alpha)]$. Namely, the player I's gain-floor and the player II's loss-ceiling have a common value. According to Definition 4.1, the α-constrained matrix game $\tilde{\boldsymbol{A}}(\alpha)$ has a value, which is also an interval, denoted by $\tilde{V}(\alpha) = [V^L(\alpha), V^R(\alpha)]$. Obviously, $\tilde{V}(\alpha) = \tilde{v}(\alpha) = \tilde{\mu}(\alpha)$, i.e., $V^L(\alpha) = v^L(\alpha) = \mu^L(\alpha)$ and $V^R(\alpha) = v^R(\alpha) = \mu^R(\alpha)$. Furthermore, it is easy to see that $\tilde{V}(\alpha)$ is the α-cut of the fuzzy value $\tilde{V}$ of the constrained matrix game $\tilde{\boldsymbol{A}}$ with payoffs of triangular fuzzy numbers, where $\tilde{V}$ is the triangular fuzzy number stated as earlier. Thus, we can draw the following conclusion, which is summarized as in Theorem 4.2.

Theorem 4.2 *For any* $\alpha \in [0, 1]$, *the* α*-constrained matrix game* $\tilde{\boldsymbol{A}}(\alpha)$ *has an interval-type* value $\tilde{V}(\alpha) = [V^L(\alpha), V^R(\alpha)]$, *whose lower and upper bounds and corresponding optimal mixed strategies for the players can be obtained through solving* Eqs. (4.12) *and* (4.13) [*or* Eqs. (4.14) *and* (4.15)], *respectively.*

Theorem 4.3 *The constrained matrix game* $\tilde{\boldsymbol{A}}$ *with payoffs of triangular fuzzy numbers always has the fuzzy value* $\tilde{V}$, *where*

$$\tilde{V} = \bigcup_{\alpha \in [0,1]} \{\alpha \otimes \tilde{V}(\alpha)\} = \bigcup_{\alpha \in [0,1]} \{\alpha \otimes [V^L(\alpha), V^R(\alpha)]\}.$$

Proof For any $\alpha \in [0, 1]$, according to Theorem 4.2, the α-constrained matrix game $\tilde{\boldsymbol{A}}(\alpha)$ has a value $\tilde{V}(\alpha) = [V^L(\alpha), V^R(\alpha)]$. Thus, it directly follows from Definition 4.2 that the constrained matrix game $\tilde{\boldsymbol{A}}$ with payoffs of triangular fuzzy numbers has a fuzzy value $\tilde{V}$. According to Eq. (2.5), we have

$$\tilde{V} = \bigcup_{\alpha \in [0,1]} \{\alpha \otimes \tilde{V}(\alpha)\} = \bigcup_{\alpha \in [0,1]} \{\alpha \otimes [V^L(\alpha), V^R(\alpha)]\}. \tag{4.18}$$

In particular, for $\alpha = 1$, according to Eqs. (4.12) and (4.13), the linear programming models are constructed as follows:

$$\max\{\boldsymbol{d}^{\mathrm{T}}\boldsymbol{x}^{R}(1)\}$$
$$\text{s.t.}\begin{cases} \boldsymbol{E}^{\mathrm{T}}\boldsymbol{x}^{R}(1) - (\boldsymbol{A}^{R}(1))^{\mathrm{T}}\boldsymbol{y}^{R}(1) \le 0 \\ \boldsymbol{B}^{\mathrm{T}}\boldsymbol{y}^{R}(1) \le \boldsymbol{c} \\ \boldsymbol{x}^{R}(1) \ge 0 \\ \boldsymbol{y}^{R}(1) \ge 0 \end{cases} \tag{4.19}$$

and

$$\max\{\boldsymbol{d}^{\mathrm{T}}\boldsymbol{x}^{L}(1)\}$$
$$\text{s.t.}\begin{cases} \boldsymbol{E}^{\mathrm{T}}\boldsymbol{x}^{L}(1) - (\boldsymbol{A}^{L}(1))^{\mathrm{T}}\boldsymbol{y}^{L}(1) \le 0 \\ \boldsymbol{B}^{\mathrm{T}}\boldsymbol{y}^{L}(1) \le \boldsymbol{c} \\ \boldsymbol{x}^{L}(1) \ge 0 \\ \boldsymbol{y}^{L}(1) \ge 0, \end{cases} \tag{4.20}$$

respectively, where $\boldsymbol{x}^{R}(1), \boldsymbol{y}^{R}(1), \boldsymbol{x}^{L}(1)$, and $\boldsymbol{y}^{L}(1)$ are vectors of decision variables, and $\boldsymbol{A}^{R}(1) = (a_{ij}^{R}(1))_{m\times n}$ and $\boldsymbol{A}^{L}(1)= (a_{ij}^{L}(1))_{m\times n}$.

Alternatively, according to Eqs. (4.14) and (4.15), the linear programming models are constructed as follows:

$$\min\{\boldsymbol{c}^{\mathrm{T}}\boldsymbol{s}^{R}(1)\}$$
$$\text{s.t.}\begin{cases} \boldsymbol{B}\boldsymbol{s}^{R}(1) - \boldsymbol{A}^{R}(1)\boldsymbol{z}^{R}(1) \ge 0 \\ \boldsymbol{E}\boldsymbol{z}^{R}(1) \ge \boldsymbol{d} \\ \boldsymbol{s}^{R}(1) \ge 0 \\ \boldsymbol{z}^{R}(1) \ge 0 \end{cases} \tag{4.21}$$

and

$$\min\{\boldsymbol{c}^{\mathrm{T}}\boldsymbol{s}^{L}(1)\}$$
$$\text{s.t.}\begin{cases} \boldsymbol{B}\boldsymbol{s}^{L}(1) - \boldsymbol{A}^{L}(1)\boldsymbol{z}^{L}(1) \ge 0 \\ \boldsymbol{E}\boldsymbol{z}^{L}(1) \ge \boldsymbol{d} \\ \boldsymbol{s}^{L}(1) \ge 0 \\ \boldsymbol{z}^{L}(1) \ge 0, \end{cases} \tag{4.22}$$

respectively, where $\boldsymbol{s}^{R}(1)$, $\boldsymbol{z}^{R}(1)$, $\boldsymbol{s}^{L}(1)$, and $\boldsymbol{z}^{L}(1)$ are vectors of decision variables.

Solving Eqs. (4.19)–(4.22) by using the simplex method of linear programming, we can obtain their optimal solutions and hereby have $v^{R}(1) = \boldsymbol{d}^{\mathrm{T}}\boldsymbol{x}^{R*}(1)$, $v^{L}(1) = \boldsymbol{d}^{\mathrm{T}}\boldsymbol{x}^{L*}(1)$, $\mu^{R}(1) = \boldsymbol{c}^{\mathrm{T}}\boldsymbol{s}^{R*}(1)$, and $\mu^{L}(1) = \boldsymbol{c}^{\mathrm{T}}\boldsymbol{s}^{L*}(1)$. It is easily derived from Eqs. (4.16) and (4.17) that $[V^L(1), V^R(1)] = [v^L(1), v^R(1)] = [\mu^L(1), \mu^R(1)]$. According to the notation of $\tilde{V} = (V^l, V^m, V^r)$, we have

$$V^m = V^L(1) = v^L(1) = V^R(1) = v^R(1) = \mu^L(1) = \mu^R(1). \quad (4.23)$$

That is to say, the mean of the fuzzy value $\tilde{V}$ of the constrained matrix game $\tilde{\boldsymbol{A}}$ with payoffs of triangular fuzzy numbers can be directly obtained by solving one of Eqs. (4.19)–(4.22). In other words, the 1-cut or Core of the fuzzy value $\tilde{V}$ is obtained as

$$\tilde{V}(1) = \mathrm{Core}(\tilde{V}) = V^m = [v^L(1), v^R(1)] = [\mu^L(1), \mu^R(1)].$$

For $\alpha = 0$, according to Eqs. (4.12) and (4.13), the linear programming models are constructed as follows:

$$\max\{\boldsymbol{d}^{\mathrm{T}}\boldsymbol{x}^R(0)\}$$
$$\text{s.t.}\begin{cases} \boldsymbol{E}^{\mathrm{T}}\boldsymbol{x}^R(0) - (\boldsymbol{A}^R(0))^{\mathrm{T}}\boldsymbol{y}^R(0) \leq 0 \\ \boldsymbol{B}^{\mathrm{T}}\boldsymbol{y}^R(0) \leq \boldsymbol{c} \\ \boldsymbol{x}^R(0) \geq 0 \\ \boldsymbol{y}^R(0) \geq 0 \end{cases} \quad (4.24)$$

and

$$\max\{\boldsymbol{d}^{\mathrm{T}}\boldsymbol{x}^L(0)\}$$
$$\text{s.t.}\begin{cases} \boldsymbol{E}^{\mathrm{T}}\boldsymbol{x}^L(0) - (\boldsymbol{A}^L(0))^{\mathrm{T}}\boldsymbol{y}^L(0) \leq 0 \\ \boldsymbol{B}^{\mathrm{T}}\boldsymbol{y}^L(0) \leq \boldsymbol{c} \\ \boldsymbol{x}^L(0) \geq 0 \\ \boldsymbol{y}^L(0) \geq 0, \end{cases} \quad (4.25)$$

respectively, where $\boldsymbol{x}^R(0), \boldsymbol{y}^R(0), \boldsymbol{x}^L(0)$, and $\boldsymbol{y}^L(0)$ are vectors of decision variables, and $\boldsymbol{A}^R(0) = (a_{ij}^R(0))_{m\times n}$ and $\boldsymbol{A}^L(0) = (a_{ij}^L(0))_{m\times n}$.

In the same way, according to Eqs. (4.14) and (4.15), the linear programming models are constructed as follows:

$$\min\{\boldsymbol{c}^{\mathrm{T}}\boldsymbol{s}^R(0)\}$$
$$\text{s.t.}\begin{cases} \boldsymbol{B}\boldsymbol{s}^R(0) - \boldsymbol{A}^R(0)\boldsymbol{z}^R(0) \geq 0 \\ \boldsymbol{E}\boldsymbol{z}^R(0) \geq \boldsymbol{d} \\ \boldsymbol{s}^R(0) \geq 0 \\ \boldsymbol{z}^R(0) \geq 0 \end{cases} \quad (4.26)$$

and

$$\min\{\boldsymbol{c}^{\mathrm{T}}\boldsymbol{s}^{L}(0)\}$$
$$\text{s.t.}\begin{cases} \boldsymbol{B}\boldsymbol{s}^{L}(0)-\boldsymbol{A}^{L}(0)\boldsymbol{z}^{L}(0)\geq 0 \\ \boldsymbol{E}\boldsymbol{z}^{L}(0)\geq \boldsymbol{d} \\ \boldsymbol{s}^{L}(0)\geq 0 \\ \boldsymbol{z}^{L}(0)\geq 0, \end{cases} \tag{4.27}$$

respectively, where $\boldsymbol{s}^{R}(0)$, $\boldsymbol{z}^{R}(0)$, $\boldsymbol{s}^{L}(0)$, and $\boldsymbol{z}^{L}(0)$ are vectors of decision variables.

Using the simplex method of linear programming, we solve Eqs. (4.24)–(4.27), respectively. Hereby, we have $v^{R}(0)=\boldsymbol{d}^{\mathrm{T}}\boldsymbol{x}^{R*}(0)$, $v^{L}(0)=\boldsymbol{d}^{\mathrm{T}}\boldsymbol{x}^{L*}(0)$, $\mu^{R}(0)=\boldsymbol{c}^{\mathrm{T}}\boldsymbol{s}^{R*}(0)$, and $\mu^{L}(0)=\boldsymbol{c}^{\mathrm{T}}\boldsymbol{s}^{L*}(0)$. It is easily derived from Eqs. (4.16) and (4.17) that $[V^{L}(0),V^{R}(0)]=[v^{L}(0),v^{R}(0)]=[\mu^{L}(0),\mu^{R}(0)]$. According to the notation of the triangular fuzzy number $\tilde{V}=(V^{l},V^{m},V^{r})$, we have

$$V^{l}=V^{L}(0)=v^{L}(0)=\mu^{L}(0),\quad V^{r}=V^{R}(0)=v^{R}(0)=\mu^{R}(0). \tag{4.28}$$

That is to say, the lower and upper bounds (or limits) of the fuzzy value $\tilde{V}$ of the constrained matrix game $\tilde{\boldsymbol{A}}$ with payoffs of triangular fuzzy numbers can be directly obtained by solving Eqs. (4.24) and (4.25) or Eqs. (4.26) and (4.27). In other words, the 0-cut or Support of the fuzzy value $\tilde{V}$ is obtained as

$$\tilde{V}(0)=\mathrm{Support}(\tilde{V})=[V^{l},V^{r}]=[v^{L}(0),v^{R}(0)]=[\mu^{L}(0),\mu^{R}(0)].$$

Theorem 4.4 *The fuzzy value of the constrained matrix game* $\tilde{\boldsymbol{A}}$ *with payoffs of triangular fuzzy numbers can be expressed as follows:*

$$\tilde{V}=\bigcup_{\alpha\in[0,1]}\{\alpha\otimes[\alpha V^{m}+(1-\alpha)V^{l},\ \alpha V^{m}+(1-\alpha)V^{r}]\}, \tag{4.29}$$

which is just about the triangular fuzzy number $\tilde{V}=(V^{l},V^{m},V^{r})$, *whose mean and lower and upper bounds (or limits) can be obtained through solving* Eqs. (4.19), (4.24) *and* (4.25) [*or one of* Eqs. (4.20)–(4.22) *and* Eqs. (4.26) *and* (4.27)], *respectively.*

Proof According to Eqs. (2.4), (4.23), and (4.28), any α-cut $\tilde{V}(\alpha)=[V^{L}(\alpha),V^{R}(\alpha)]$ of the fuzzy value $\tilde{V}$ of the constrained matrix game $\tilde{\boldsymbol{A}}$ with payoffs of triangular fuzzy numbers can be obtained as follows:

$$\tilde{V}(\alpha)=\alpha\tilde{V}(1)+(1-\alpha)\tilde{V}(0)=[\alpha V^{m}+(1-\alpha)V^{l},\alpha V^{m}+(1-\alpha)V^{r}].$$

Combining with Theorem 4.3, the fuzzy value $\tilde{V}$ can be expressed as follows:

$$\tilde{V} = \bigcup_{\alpha \in [0,1]} \{\alpha \otimes \tilde{V}(\alpha)\} = \bigcup_{\alpha \in [0,1]} \{\alpha \otimes [\alpha V^m + (1-\alpha)V^l, \alpha V^m + (1-\alpha)V^r]\},$$

which directly implies that the membership function of $\tilde{V}$ is given as follows:

$$\mu_{\tilde{V}}(x) = \max\{\alpha | \alpha V^m + (1-\alpha)V^l \leq x \leq \alpha V^m + (1-\alpha)V^r\}$$
$$= \begin{cases} \frac{x-V^l}{V^m-V^l} & \text{if} \quad V^l \leq x < V^m \\ 1 & \text{if} \quad x = V^m \\ \frac{V^r-x}{V^r-V^m} & \text{if} \quad V^m < x \leq V^r \\ 0 & \text{else.} \end{cases}$$

Therefore, the fuzzy value $\tilde{V}$ of the constrained matrix game $\tilde{\boldsymbol{A}}$ with payoffs of triangular fuzzy numbers is just about the triangular fuzzy number (V^l, V^m, V^r). Thus, we have completed the proof of Theorem 4.4.

Theorem 4.4 shows that the fuzzy value $\tilde{V}$ of any constrained matrix game $\tilde{\boldsymbol{A}}$ with payoffs of triangular fuzzy numbers is a triangular fuzzy number, which can be explicitly obtained through solving the derived three auxiliary linear programming models with all data taken from only the 1-cut and 0-cut of the payoffs represented by triangular fuzzy numbers.

4.3.3 Algorithm of Linear Programming Method of Constrained Matrix Games with Payoffs of Triangular Fuzzy Numbers

From the aforementioned discussion, the process and algorithm for solving constrained matrix games with payoffs of triangular fuzzy numbers are summarized as follows.

Step 1 Identify players, denoted by I and II;

Step 2 Identify pure strategies of the players I and II, denoted the sets of pure strategies by $S_1 = \{\delta_1, \delta_2, \ldots, \delta_m\}$ and $S_2 = \{\beta_1, \beta_2, \ldots, \beta_n\}$, respectively;

Step 3 Identify constraint conditions of strategies for the players I and II, denoted the constrained sets of strategies by Y and Z, respectively;

Step 4 Pool opinions of outcomes for the players I and II and estimate the player I's payoffs expressed with triangular fuzzy numbers $\tilde{a}_{ij} = (a_{ij}^l, a_{ij}^m, a_{ij}^r)$ $(i = 1, 2, \ldots, m;\ j = 1, 2, \ldots, n)$ and hereby construct the payoff matrix $\tilde{\boldsymbol{A}} = (\tilde{a}_{ij})_{m \times n}$;

Step 5 Construct and solve the linear programming models according to Eqs. (4.12) and (4.13) [or Eqs. (4.14) and (4.15)] and hereby obtain the value $\tilde{V}(\alpha)$ and corresponding optimal mixed strategies for the players in any α-constrained matrix game $\tilde{\boldsymbol{A}}(\alpha)$, where $\alpha \in [0,1]$;

Step 6 Construct and solve the linear programming model according to one of Eqs. (4.19)–(4.22) and hereby obtain V^m;

Step 7 Construct and solve the two linear programming problems according to Eqs. (4.24) and (4.25) [or Eqs. (4.26) and (4.27)] and hereby obtain V^l and V^r;

Step 8 Construct the fuzzy value $\tilde{V}$ of the constrained matrix game $\tilde{\boldsymbol{A}}$ with payoffs of triangular fuzzy numbers according to the obtained values V^m, V^l , and V^r. Namely, we can explicitly obtain the fuzzy value $\tilde{V}$, which is expressed as the triangular fuzzy number $\tilde{V} = (V^l, V^m, V^r)$.

4.3.4 Real Example Analysis of Market Share Problems with Payoffs of Triangular Fuzzy Numbers

In this subsection, let us consider the following numerical example, which is adopted from the example of the market share game problem in Sect. 3.3.3. More precisely, we briefly describe the situation of this example as follows.

Two business companies p_1 and p_2 plan to enhance the market share of an electronic product in a targeted market under the circumstance that the demand amount of the electronic product in the targeted market basically is fixed. They are considering about two options (i.e., pure strategies) to increase the market share: improving technology (δ_1), advertisement (δ_2). The company p_1 needs the funds 80 and 50 (million Yuan) when it takes the pure strategies δ_1 and δ_2, respectively. However, due to the lack of the funds, the company p_1 only has 60 (million Yuan), i.e., the mixed strategies of the company p_1 must satisfy the constraint condition: $80y_1 + 50y_2 \le 60$. Similarly, the mixed strategies of the company p_2 must satisfy the constraint condition: $40z_1 + 70z_2 \le 50$ or $-40z_1 - 70z_2 \ge -50$. Due to a lack of information or imprecision of the available information, the managers of the two companies usually are not able to exactly forecast the sales amount of the companies' product. Thus, triangular fuzzy numbers are used to represent the sales amount of the product from both companies' perspectives. Therefore, the above problem may be regarded as a constrained matrix game with payoffs of triangular fuzzy numbers. Namely, the companies p_1 and p_2 are regarded as the players I and II, respectively. The constrained sets of strategies for the players I and II are expressed as follows:

$$Y_0 = \{\boldsymbol{y} | 80y_1 + 50y_2 \le 60, y_1 + y_2 \le 1, -y_1 - y_2 \le -1, y_1 \ge 0, y_2 \ge 0\}$$

and

$$Z_0 = \{z| -40z_1 - 70z_2 \geq -50, z_1 + z_2 \leq 1, -z_1 - z_2 \leq -1, z_1 \geq 0, z_2 \geq 0\},$$

respectively. Let us consider the specific constrained matrix game $\tilde{A}_0$ with payoffs of triangular fuzzy numbers, where the fuzzy payoff matrix of the player I is given as follows:

$$\tilde{\boldsymbol{A}}_0 = \begin{matrix} & \delta_1 & \delta_2 \\ \begin{matrix}\delta_1\\ \delta_2\end{matrix} & \multicolumn{2}{c}{\begin{pmatrix} (27,29,35) & (-25,-19,-17) \\ (-11,-10,-5) & (35,40,41) \end{pmatrix}} \end{matrix},$$

where the element $(27, 29, 35)$ in the fuzzy payoff matrix $\tilde{\boldsymbol{A}}_0$ is a triangular fuzzy number, which indicates that the sales amount of the product for the company p_1 is between 27 and 35 when the companies p_1 and p_2 use the pure strategy δ_1 (improving technology) simultaneously. Other elements (i.e., triangular fuzzy numbers) in the fuzzy payoff matrix $\tilde{\boldsymbol{A}}_0$ can be explained similarly.

The coefficient matrices and vectors of the constrained sets Y_0 and Z_0 of mixed strategies for the companies p_1 and p_2 (i.e., the players I and II) are obtained as follows:

$$\boldsymbol{B}_0 = \begin{pmatrix} 80 & 1 & -1 \\ 50 & 1 & -1 \end{pmatrix},$$

$$\boldsymbol{E}_0^{\mathrm{T}} = \begin{pmatrix} -40 & 1 & -1 \\ -70 & 1 & 1 \end{pmatrix}$$

and

$$\boldsymbol{c} = (60, 1, -1)^{\mathrm{T}}, \quad \boldsymbol{d} = (-50, 1, -1)^{\mathrm{T}},$$

respectively.

1. Computational results obtained by the proposed linear programming method

According to Eq. (4.19), the linear programming model can be constructed as follows:

$$\max\{-50x_1^R(1) + x_2^R(1) - x_3^R(1)\}$$
$$\text{s.t.} \begin{cases} -40x_1^R(1) + x_2^R(1) - x_3^R(1) - 29y_1^R(1) + 10y_2^R(1) \leq 0 \\ -70x_1^R(1) + x_2^R(1) - x_3^R(1) + 19y_1^R(1) - 40y_2^R(1) \leq 0 \\ 80y_1^R(1) + 50y_2^R(1) \leq 60 \\ y_1^R(1) + y_2^R(1) \leq 1 \\ -y_1^R(1) - y_2^R(1) \leq -1 \\ x_1^R(1) \geq 0,\ x_2^R(1) \geq 0,\ x_3^R(1) \geq 0,\ y_1^R(1) \geq 0,\ y_2^R(1) \geq 0. \end{cases} \tag{4.30}$$

Solving Eq. (4.30) by using the simplex method of linear programming [4], we obtain its optimal solution $(\boldsymbol{x}^{R*}(1), \boldsymbol{y}^{R*}(1))$, where $\boldsymbol{y}^{R*}(1) = (1/3, 2/3)^{\mathrm{T}}$ and $\boldsymbol{x}^{R*}(1) = (0, 3, 0)^{\mathrm{T}}$. Therefore, the upper bound $v^R(1)$ of the gain-floor for the company p_1 and the optimal mixed strategy $\boldsymbol{y}^{R*}(1)$ are obtained as $v^R(1) = \boldsymbol{d}^{\mathrm{T}}\boldsymbol{x}^{R*}(1) = 3$ and $\boldsymbol{y}^{R*}(1) = (1/3, 2/3)^{\mathrm{T}}$, respectively. Thus, the mean of the fuzzy value of the constrained matrix game $\tilde{\boldsymbol{A}}_0$ with payoffs of triangular fuzzy numbers and corresponding optimal mixed strategy are $V^m = v^R(1) = 3$ and $\boldsymbol{y}^{m*} = \boldsymbol{y}^{R*}(1) = (1/3, 2/3)^{\mathrm{T}}$, respectively.

According to Eq. (4.24), the linear programming model can be constructed as follows:

$$\max\{-50x_1^R(0) + x_2^R(0) - x_3^R(0)\}$$
$$\text{s.t.}\begin{cases} -40x_1^R(0) + x_2^R(0) - x_3^R(0) - 35y_1^R(0) + 5y_2^R(0) \leq 0 \\ -70x_1^R(0) + x_2^R(0) - x_3^R(0) + 17y_1^R(0) - 41y_2^R \leq 0 \\ 80y_1^R(0) + 50y_2^R(0) \leq 60 \\ y_1^R(0) + y_2^R(0) \leq 1 \\ -y_1^R(0) - y_2^R(0) \leq -1 \\ x_1^R(0) \geq 0,\ x_2^R(0) \geq 0,\ x_3^R(0) \geq 0,\ y_1^R(0) \geq 0,\ y_2^R(0) \geq 0. \end{cases} \tag{4.31}$$

Solving Eq. (4.31) by using the simplex method of linear programming, we obtain its optimal solution $(\boldsymbol{x}^{R*}(0), \boldsymbol{y}^{R*}(0))$, where $\boldsymbol{y}^{R*}(0) = (1/3, 2/3)^{\mathrm{T}}$ and $\boldsymbol{x}^{R*}(0) = (0, 8.3333, 0)^{\mathrm{T}}$. Therefore, the upper bound (or limit) $v^R(0)$ of the gain-floor for the company p_1 and the optimal mixed strategy $\boldsymbol{y}^{R*}(0)$ are $v^R(0) = \boldsymbol{d}^{\mathrm{T}}\boldsymbol{x}^{R*}(0) = 8.3333$ and $\boldsymbol{y}^{R*}(0) = (1/3, 2/3)^{\mathrm{T}}$, respectively. Then, the upper bound of the fuzzy value of the constrained matrix game $\tilde{\boldsymbol{A}}_0$ with payoffs of triangular fuzzy numbers and corresponding optimal mixed strategy are $V^r = v^R(0) = 8.3333$ and $\boldsymbol{y}^{r*} = \boldsymbol{y}^{R*}(0) = (1/3, 2/3)^{\mathrm{T}}$, respectively.

Analogously, according to Eq. (4.25), the linear programming model can be constructed as follows:

$$\max\{-50x_1^L(0) + x_2^L(0) - x_3^L(0)\}$$
$$\text{s.t.}\begin{cases} 40\ x_1^L(0) + x_2^L(0) - x_3^L(0) - 27y_1^L(0) + 11y_2^L(0) \leq 0 \\ -70x_1^L(0) + x_2^L(0) - x_3^L(0) + 25y_1^L(0) - 35y_2^L(0) \leq 0 \\ 80y_1^L(0) + 50y_2^L(0) \leq 60 \\ y_1^L(0) + y_2^L(0) \leq 1 \\ -y_1^L(0) - y_2^L(0) \leq -1 \\ x_1^L(0) \geq 0,\ x_2^L(0) \geq 0,\ x_3^L(0) \geq 0,\ y_1^L(0) \geq 0,\ y_2^L(0) \geq 0. \end{cases} \tag{4.32}$$

Solving Eq. (4.32) by using the simplex method of linear programming, we obtain its optimal solution $(\boldsymbol{x}^{L*}(0), \boldsymbol{y}^{L*}(0))$, where $\boldsymbol{y}^{L*}(0) = (1/3, 2/3)^{\mathrm{T}}$ and $\boldsymbol{x}^{L*}(0) = (0, 1.6667, 0)^{\mathrm{T}}$. Therefore, the lower bound (or limit) $v^L(0)$ of the gain-floor for the company p_1 and the optimal mixed strategy $\boldsymbol{y}^{L*}(0)$ are $v^L(0) = \boldsymbol{d}^{\mathrm{T}}\boldsymbol{x}^{L*}(0) = 1.6667$ and $\boldsymbol{y}^{L*}(0) = (1/3, 2/3)^{\mathrm{T}}$, respectively. Hence, we obtain the lower bound $V^l = v^L(0) = 1.6667$ of the fuzzy value of the constrained matrix game $\tilde{\boldsymbol{A}}_{\boldsymbol{0}}$ with payoffs of triangular fuzzy numbers and corresponding optimal mixed strategy $\boldsymbol{y}^{l*} = \boldsymbol{y}^{L*}(0) = (1/3, 2/3)^{\mathrm{T}}$. Thus, the fuzzy value of the constrained matrix game $\tilde{\boldsymbol{A}}_{\boldsymbol{0}}$ with payoffs of triangular fuzzy numbers is a triangular fuzzy number $\tilde{V} = (V^l, V^m, V^r) = (1.6667, 3, 8.3333)$, whose membership function is given as follows:

$$\mu_{\tilde{V}}(x) = \begin{cases} \frac{x-1.6667}{1.3333} & \text{if } 1.6667 \le x < 3 \\ 1 & \text{if } x = 3 \\ \frac{8.3333-x}{5.3333} & \text{if } 3 < x \le 8.3333 \\ 0 & \text{else,} \end{cases}$$

depicted as in Fig. 4.2.

Alternatively, according to Eq. (4.21), the linear programming model can be constructed as follows:

$$\min\{60s_1^R(1) + s_2^R(1) - s_3^R(1)\}$$
$$\text{s.t.} \begin{cases} 80s_1^R(1) + s_2^R(1) - s_3^R(1) - 29z_1^R(1) + 19z_2^R(1) \ge 0 \\ 50s_1^R(1) + s_2^R(1) - s_3^R(1) + 10z_1^R(1) - 40z_2^R(1) \ge 0c \\ -40z_1^R(1) - 70z_2^R(1) \ge -50 \\ z_1^R(1) + z_2^R(1) \ge 1 \\ -z_1^R(1) - z_2^R(1) \ge -1 \\ s_1^R(1) \ge 0, s_2^R(1) \ge 0, s_3^R(1) \ge 0, z_1^R(1) \ge 0, z_2^R(1) \ge 0. \end{cases} \quad (4.33)$$

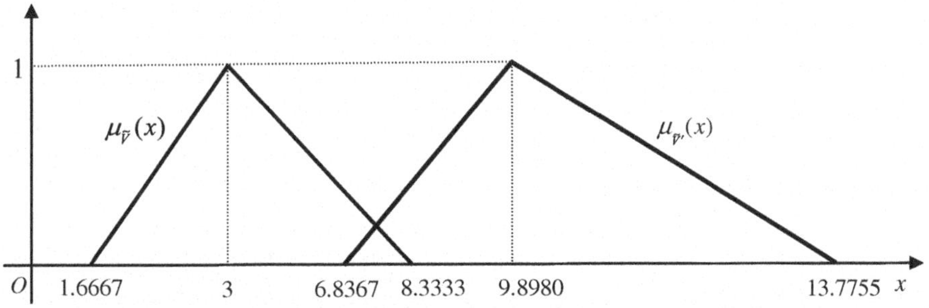

Fig. 4.2 The fuzzy values $\tilde{V}$ and $\tilde{V}'$

Solving Eq. (4.33) by using the simplex method of linear programming, we obtain its optimal solution $(\boldsymbol{s}^{R*}(1), \boldsymbol{z}^{R*}(1))$, where $\boldsymbol{z}^{R*}(1) = (1, 0)^{\mathrm{T}}$ and $\boldsymbol{s}^{R*}(1) = (1.3, 0, 75.0)^{\mathrm{T}}$. Therefore, the upper bound $\mu^R(1)$ of the loss-ceiling for the company p_2 and the optimal mixed strategy $\boldsymbol{z}^{R*}(1)$ are $\mu^R(1) = \boldsymbol{d}^{\mathrm{T}}\boldsymbol{s}^{R*}(1) = 3$ and $\boldsymbol{z}^{R*}(1) = (1, 0)^{\mathrm{T}}$, respectively.

Likewise, according to Eq. (4.26), the linear programming model can be constructed as follows:

$$
\min\{60s_1^R(0) + s_2^R(0) - s_3^R(0)\}
$$
$$
\text{s.t.}\begin{cases}
80s_1^R(0) + s_2^R(0) - s_3^R(0) - 35z_1^R(0) + 17z_2^R(0) \geq 0 \\
50s_1^R(0) + s_2^R(0) - s_3^R(0) + 5z_1^R(0) - 41z_2^R(0) \geq 0 \\
-40z_1^R(0) - 70z_2^R(0) \geq -50 \\
z_1^R(0) + z_2^R(0) \geq 1 \\
-z_1^R(0) - z_2^R(0) \geq -1 \\
s_1^R(0) \geq 0, s_2^R(0) \geq 0, s_3^R(0) \geq 0, z_1^R(0) \geq 0, z_2^R(0) \geq 0.
\end{cases} \tag{4.34}
$$

Solving Eq. (4.34) by using the simplex method of linear programming, we obtain its optimal solution $(\boldsymbol{s}^{R*}(0), \boldsymbol{z}^{R*}(0))$, where $\boldsymbol{z}^{R*}(0) = (1, 0)^{\mathrm{T}}$ and $\boldsymbol{s}^{R*}(0) = (1.3333, 0, 71.6667)^{\mathrm{T}}$, respectively. Therefore, the upper bound (or limit) $\mu^R(0)$ of the loss-ceiling for the company p_2 and the optimal mixed strategy $\boldsymbol{z}^{R*}(0)$ are $\mu^R(0) = \boldsymbol{d}^{\mathrm{T}}\boldsymbol{s}^{R*}(0) = 8.3333$ and $\boldsymbol{z}^{R*}(0) = (1, 0)^{\mathrm{T}}$, respectively.

According to Eq. (4.27), the linear programming model can be constructed as follows:

$$
\min\{60s_1^L(0) + s_2^L(0) - s_3^L(0)\}
$$
$$
\text{s.t.}\begin{cases}
80s_1^L(0) + s_2^L(0) - s_3^L(0) - 27z_1^L(0) + 25z_2^L(0) \geq 0 \\
50s_1^L(0) + s_2^L(0) - s_3^L(0) + 11z_1^L(0) - 35z_2^L(0) \geq 0 \\
-40z_1^L(0) - 70z_2^L(0) \geq -50 \\
z_1^L(0) + z_2^L(0) \geq 1 \\
-z_1^L(0) - z_2^L(0) \geq -1 \\
s_1^L(0) \geq 0, s_2^L(0) \geq 0, s_3^L(0) \geq 0, z_1^L(0) \geq 0, z_2^L(0) \geq 0.
\end{cases} \tag{4.35}
$$

Solving Eq. (4.35) by using the simplex method of linear programming, we obtain its optimal solution $(\boldsymbol{s}^{L*}(0), \boldsymbol{z}^{L*}(0))$, where $\boldsymbol{z}^{L*}(0) = (1, 0)^{\mathrm{T}}$ and $\boldsymbol{s}^{L*}(0) = (1.2667, 0, 74.3333)^{\mathrm{T}}$. Therefore, the lower bound (or limit) $\mu^L(0)$ of the loss-ceiling for the company p_2 and the optimal mixed strategy $\boldsymbol{z}^{L*}(0)$ are $\mu^L(0) = \boldsymbol{d}^{\mathrm{T}}\boldsymbol{s}^{L*}(0) = 1.6667$ and $\boldsymbol{z}^{L*}(0) = (1, 0)^{\mathrm{T}}$, respectively.

From the above computational results, obviously, the fuzzy value of the constrained matrix game $\tilde{\boldsymbol{A}}_0$ with payoffs of triangular fuzzy numbers obtained by using

Eqs. (4.21), (4.26), and (4.27) is the same as that obtained by using Eqs. (4.19), (4.24), and (4.25), i.e., the triangular fuzzy number

$$\tilde{V} = (V^l, V^m, V^r) = (1.6667, 3, 8.3333).$$

2. Computational results without considering constrained strategies

If the companies p_1 and p_2 have sufficient funds, i.e., both companies do not take into account the constraint conditions of the strategies, then the above market share problem may be regarded as an unconstrained matrix game $\tilde{A}'_0$ with payoffs of triangular fuzzy numbers. Thus, using the linear programming method for solving interval-valued matrix games developed by Li [10], we construct the linear programming model as follows:

$$\min\{x_1'^R(1) + x_2'^R(1)\}$$
$$\text{s.t.} \begin{cases} 29x_1'^R(1) - 10x_2'^R(1) \geq 1 \\ -19x_1'^R(1) + 40x_2'^R(1) \geq 1 \\ x_1'^R(1) \geq 0,\ x_2'^R(1) \geq 0, \end{cases} \tag{4.36}$$

where $x_1'^R(1)$ and $x_2'^R(1)$ are decision variables.

Solving Eq. (4.36) by using the simplex method of linear programming, we obtain its optimal solution $\boldsymbol{x}'^{R*}(1) = (x_1'^{R*}(1), x_2'^{R*}(1))^{\mathrm{T}} = (0.0515, 0.0495)^{\mathrm{T}}$. By the method given by Li [10], we obtain the mean v'^m of the gain-floor of the unconstrained matrix game $\tilde{A}'_0$ with payoffs of triangular fuzzy numbers and the optimal mixed strategy $\boldsymbol{y}'^{m*}$ for the player I (i.e., the company p_1), where

$$v'^m = v'^R(1) = \frac{1}{x_1'^{R*}(1) + x_2'^{R*}(1)} = 9.8980,$$

$$y_1'^{m*} = y_1'^{R*}(1) = \frac{x_1'^{R*}(1)}{x_1'^{R*}(1) + x_2'^{R*}(1)} = 0.5102$$

and

$$y_2'^{m*} = y_2'^{R*}(1) = \frac{x_2'^{R*}(1)}{x_1'^{R*}(1) + x_2'^{R*}(1)} = 0.4898.$$

Analogously, the two linear programming models are constructed as follows:

$$\min\{x_1'^R(0) + x_2'^R(0)\}$$
$$\text{s.t.} \begin{cases} 35x_1'^R(0) - 5x_2'^R(0) \geq 1 \\ -17x_1'^R(0) + 41x_2'^R(0) \geq 1 \\ x_1'^R(0) \geq 0, x_2'^R(0) \geq 0 \end{cases} \tag{4.37}$$

and

$$\min\{x_1'^L(0)+x_2'^L(0)\}$$
$$\text{s.t.}\begin{cases} 27x_1'^L(0)-11x_2'^L(0)\geq 1 \\ -25x_1'^L(0)+35x_2'^L(0)\geq 1 \\ x_1'^L(0)\geq 0,\ x_2'^L(0)\geq 0, \end{cases} \quad (4.38)$$

where $x_1'^R(0)$, $x_2'^R(0)$, $x_1'^L(0)$, and $x_2'^L(0)$ are decision variables.

Solving Eqs. (4.37) and (4.38) by using the simplex method of linear programming, we can obtain their optimal solutions and hereby obtain the lower and upper bounds of the gain-floor of the unconstrained matrix game $\tilde{A}_0'$ with payoffs of triangular fuzzy numbers and corresponding optimal mixed strategies for the player I, where

$$v'^r=13.7755,\quad \boldsymbol{y}'^{r*}=(0.4694,0.5306)^{\mathrm{T}}$$

and

$$v'^l=6.8367,\quad \boldsymbol{y}'^{l*}=(0.4694,0.5306)^{\mathrm{T}}.$$

Therefore, the gain-floor of the unconstrained matrix game $\tilde{A}_0'$ with payoffs of triangular fuzzy numbers for the company p_1 is a triangular fuzzy number $\tilde{v}'=(v'^l,v'^m,v'^r)=(6.8367,9.8980,13.7755)$, whose membership function is given as follows:

$$\mu_{\tilde{v}'}(x)=\begin{cases} \frac{x-6.8367}{3.0613} & \text{if} \quad 6.8367\leq x<9.8980 \\ 1 & \text{if} \quad x=9.8980 \\ \frac{13.7755-x}{3.8775} & \text{if} \quad 9.8980<x\leq 13.7755 \\ 0 & \text{else.} \end{cases}$$

In the same way, using the method given by Li [10], we have

$$\mu'^m=9.8980,\quad z'^{m*}=(0.6020,0.3980)^{\mathrm{T}},$$
$$\mu'^r=13.7755,\quad z'^{r*}=(0.5918,0.4082)^{\mathrm{T}}$$

and

$$\mu'^l=6.8367,\quad z'^{l*}=(0.6122,0.3878)^{\mathrm{T}}.$$

Then, the loss-ceiling of the unconstrained matrix game $\tilde{A}_0'$ with payoffs of triangular fuzzy numbers for the company p_2 is a triangular fuzzy number

$$\tilde{\mu}' = (\mu'^l, \mu'^m, \mu'^r) = (6.8367, 9.8980, 13.7755).$$

Thus, the companies p_1 and p_2 have a common fuzzy value, i.e., $\tilde{v}' = \tilde{\mu}' = (6.8367, 9.8980, 13.7755)$. Hereby, the unconstrained matrix game $\tilde{A}_0'$ with payoffs of triangular fuzzy numbers has a fuzzy value, i.e., the triangular fuzzy number $\tilde{V}' = (6.8367, 9.8980, 13.7755)$, depicted as in Fig. 4.2.

It is easy to see from Fig. 4.2 that the fuzzy value $\tilde{V}' = (6.8367, 9.8980, 13.7755)$ and companies' optimal mixed strategies in the unconstrained matrix game $\tilde{A}_0'$ with payoffs of triangular fuzzy numbers are different from the fuzzy value $\tilde{V} = (1.6667, 3, 8.3333)$ and optimal mixed strategies in the constrained matrix game $\tilde{A}_0$ with payoffs of triangular fuzzy numbers. Moreover, obviously, $\tilde{V}' = (6.8367, 9.8980, 13.7755)$ is larger than $\tilde{V} = (1.6667, 3, 8.3333)$. These conclusions are accordance with the actual situation as expected. On the other hand, it is shown that it is necessary to consider the constraint conditions of strategies in real situations.

References

1. Zadeh LA (1975) The concept of a linguistic variable and its applications to approximate reasoning, Part I. Inf Sci 8:199–249
2. Li D-F, Cheng C-T (2002) Fuzzy multiobjective programming methods for fuzzy constrained matrix games with fuzzy numbers. Int J Uncertainty Fuzziness Knowl Based Syst 10(4):385–400
3. Li D-F (1999) Fuzzy constrained matrix games with fuzzy payoffs. J Fuzzy Math 7(4):873–880
4. Owen G (1982) Game theory, 2nd edn. Academic Press, New York
5. Li D-F (2003) Fuzzy multiobjective many-person decision makings and games. National Defense Industry Press, Beijing (in Chinese)
6. Ramik J, Rimanek J (1985) Inequality relation between fuzzy numbers and its use in fuzzy optimization. Fuzzy Sets Syst 16:123–138
7. Zimmermann H-J (1991) Fuzzy set theory and its application, 2nd edn. Kluwer Academic Publishers, Dordrecht
8. Li D-F (2005) An approach to fuzzy multiattribute decision making under uncertainty. Inf Sci 169(1–2):97–112
9. Lai Y-J, Hwang C-L (1992) A new approach to some possibilistic linear programming problems. Fuzzy Sets Syst 49:121–133
10. Li D-F (2011) Linear programming approach to solve interval-valued matrix games. Omega: Int J Manag Sci 39(6):655–666
11. Moore RE (1979) Method and application of interval analysis. SIAM, Philadelphia

The manufacturer's authorised representative in the EU is Springer Nature Customer Service Centre GmbH, Europaplatz 3, 69115 Heidelberg, Germany. If you have any concerns regarding our products, please contact ProductSafety@springernature.com

Printed and bound by CPI Group (UK) Ltd, Croydon, CR0 4YY
23/03/2026
02076380-0013